POPULAR SCIENCE

THE FUTURE NOW

FYI

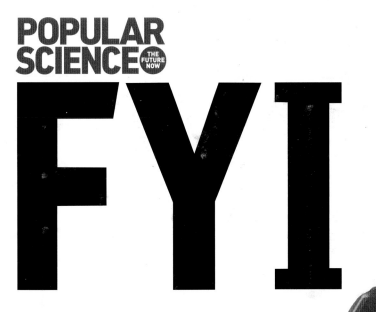

FYI

WHAT DOES SPACE SMELL LIKE?
CAN A PAPER CUT BE DEADLY?
JUST HOW OLD IS DIRT?

229 CURIOUS QUESTIONS ANSWERED
BY THE WORLD'S SMARTEST PEOPLE

weldon**owen**

INTRODUCTION

THE FINAL FRONTIER

CONTENTS

HUMAN EXTREMES

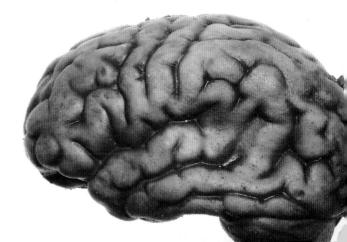

THE ANIMAL WORLD

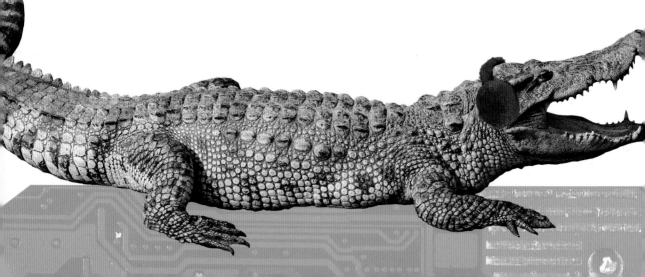

HERE ON EARTH

EVERYDAY LIFE

WONDERS OF MEDICINE

A WORD FROM FYI EDITOR BJORN CAREY

▶ Hello, dear reader! Thank you for purchasing *FYI*, the best of *Popular Science* magazine's "FYI" column, which consults top experts to answer your toughest science questions. You have made a wise investment, as the articles within will open your brain to a universe of uncommon knowledge.

You will learn some very useful life skills in these pages. Should your child float away while you're vacationing in an orbital space hotel, you'll know the best way to attempt a rescue (009). If you suspect that you share your home with some creepy paranormal tenants, you'll know how to smoke them out (145). If you ever need to recycle a nuclear warhead, you'll be able to identify the most dangerous part of the bomb (137). And if you find yourself in the middle of nowhere suffering from a terrible case of jock itch, you'll know a simple recipe for cooking up an all-natural cure (066). Handy things, all of these, but the true reward of the following pages will be familiarizing yourself with the *FYI* thinking process: Not all questions have an answer, but the search can enlighten.

Formulating questions and then relentlessly searching for answers is the foundation of good science. Consider the story of Percy Spencer. In 1945, Spencer, a self-taught engineer working for the Raytheon Corporation, was busy building magnetrons for a radar device. One day in the lab, he noticed that the chocolate and peanut candy bar in his pocket melted. "That's weird," he thought, "there's no obvious heat source to ruin my afternoon treat." Poking around his workstation, he realized that the magnetron was emitting microwaves, and that these invisible streams of energy had excited the atoms in his chocolate and melted the bar. Hypothesis in hand, he placed—get this—a bowl of corn kernels in front of the device. And they popped! Then he fired the microwaves at an egg. It exploded!

By October of that year he had verified the mechanism and Raytheon filed a patent for the microwave oven. The accidental discovery of one of the most ubiquitous kitchen appliances was made possible simply because Spencer was curious.

Indeed, throughout history, science has rewarded the curious. Decades before Alexander Fleming milked antibiotics out of *Penicillium* fungi, someone looked at that same blue mold and thought, "Hey. I wonder if rubbing this on my festering wound will fight off infection and help it heal more quickly?" And it did.

Now, hunting down the reasons why one should not eat canned cat food (054) won't yield anything as useful as penicillin. But I hope that the persistent manner in which we have chased down answers for never-before-asked questions will inspire in you, dear reader, the type of curiosity that will lead to the next great scientific achievement.

And while I hope that *FYI* inspires, I also hope that you enjoy the heck out of it. We've enthusiastically logged hundreds of hours consulting the the best and brightest minds as we dug up satisfying answers. (You may or may not be shocked to learn that more than half a dozen bioethicists refused to speak on the ethical dilemmas of eating cloned human flesh.) I also hope that reading this will quiet your urges to eat a star, dance inside a particle accelerator, or try to get a DeLorean airborne. As you will soon find out, these are terrible ideas. Might I suggest making your own gunpowder instead? Onward!

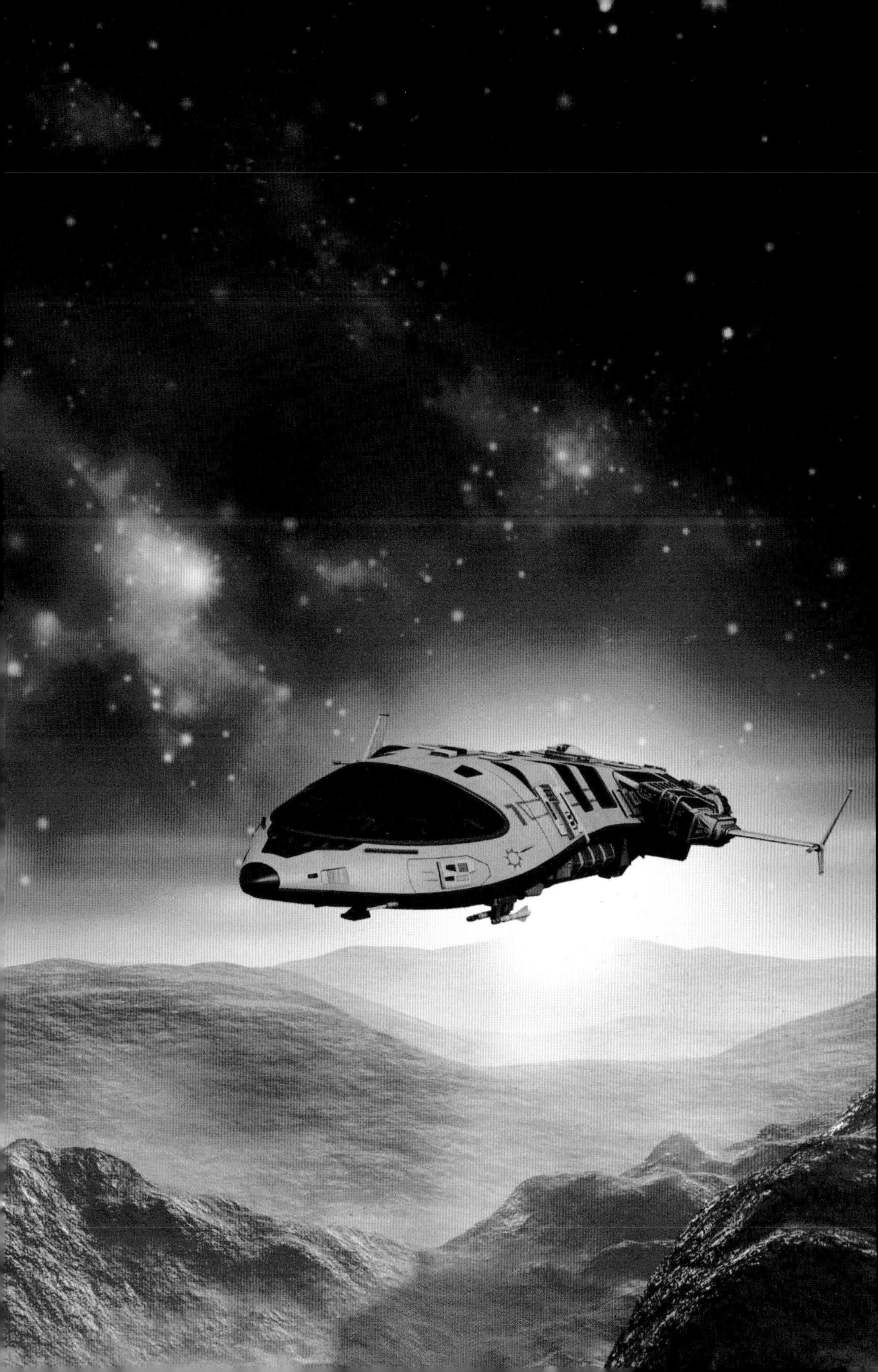

THE FINAL FRONTIER

001

WHAT WOULD BE THE BEST JOB IN A MARS COLONY?

▶ **OPPORTUNITY KNOCKS**

Looking for a big career change? Robert Zubrin, president of the Mars Society and author of *How to Live on Mars*, offers some suggestions on how to get ahead on the Red Planet.

"You've just landed on Mars and are looking for a fresh start. Sure, that job selling respirators at the local space-hardware store sounds cozy, but it's a dead-end career. Mars will be rife with opportunity; you just have to figure out how to tap it. So here's the secret: Go into construction. You'll learn useful skills and get to spend time out on the surface, where the real action is. Explore the landscape on coffee breaks. All you need to do is stumble upon a nice deposit of precious material, like platinum or deuterium (a hydrogen isotope that could fuel fusion reactors), and you'll have it made.

"Next, buddy up with the engineers working to terraform the Martian hillsides. It's their job to turn all that red dust into Earthlike soil that can support robust vegetation and seed the atmosphere so it will rain and form lakes and oceans. Figure out where future beachfront property will be, buy it, and auction off lots to the highest bidder.

"Of course, this prosperous career path has its risks. You'll be outdoors a lot, and Mars's atmosphere is pretty thin, so cosmic radiation could fry your DNA. Things could fall on you at construction sites. And you'd probably go prospecting alone (why split the profits?), so no one could help you if you got lost or fell into a crater. You could play it safe in the colony, working at the Spacemart. But you're on Mars—take a chance!"

IF MY SPACE SUIT TORE, WOULD MY HEAD EXPLODE? 002

IN SPACE, NO ONE CAN HEAR YOU SCREAM

Hollywood makes death in space seem really gruesome, but the good news is your head won't actually explode if your space suit cracks. The bad news? Your blood might boil, and you'd still probably die. Let us explain.

First, let's get those special effects–enhanced deaths out of the way. The exploding heads in the movies come from a misinterpretation of basic physics: Pressurized objects expand under lower pressure, so your body, which is used to life at 1 atmosphere of pressure, might reasonably be expected to expand (violently) when it's exposed to the zero-pressure vacuum of space. In reality, the pressure difference just isn't great enough for that type of reaction. (Your skin, however, would swell to about twice its normal size. This is painful, but survivable. Once you got back to Earth, it would snap back to its original shape.)

The boiling blood is a little trickier—scientists are divided on this issue. It's well established that the boiling points of liquids decrease as pressure drops. At zero pressure, your 98.6°F (37°C) body should provide enough heat to boil blood, which is the main support for the "blood boils" campaign. Scientists on the other side present a persuasive counterargument, however: Since the circulatory system is a closed loop, a beating heart provides constant pressure, and your blood vessels are elastic, your blood would stay relatively compressed. There would still be a drop in pressure, but not enough for your blood to boil at body temperature. A third contingent says that the blood might only *seem* to boil as dissolved oxygen and nitrogen gasified out of it. (Saliva in your mouth *would* actually boil, at least according to one NASA astronaut who was accidentally exposed to vacuum during a 1967 training exercise.)

003 WHAT WOULD REALLY HAPPEN, THEN?

INTO THIN AIR

In reality, you'd likely die of asphyxiation as oxygen leaked from your suit. This isn't just theoretical, unfortunately—in 1971, the crew of the *Soyuz 11* died of asphyxiation before reentry due to a valve malfunction. When officials found the capsule, the bodies inside showed no signs of trauma. It was only after autopsies were performed that officials concluded that the cosmonauts died of a lack of oxygen.

If you find yourself in this situation, don't hold your breath. If your lungs are full of air, the difference in pressure between them and space will cause explosive decompression—decompression because the air rapidly expands, and explosive because, well, you can probably guess. Explosive decompression can burst your lungs, which, while not as graphic as Hollywood would have you believe, is still quite unpleasant.

004

WHAT MODIFICATIONS WOULD I NEED TO MAKE TO MY CAR SO I COULD DRIVE IT ON THE MOON?

▶ MAY INVALIDATE WARRANTY

When the Apollo astronauts drove around on the moon, they had to settle for a little buggy. But if you want to tour the Sea of Tranquility in a Ferrari or the family SUV, well, you're looking at more than a few weekends under the hood.

"Your average car faces several major problems on the moon," says Brian Wilcox, who heads the development of NASA's future manned rover, called the Athlete. Chief among these is the small matter of combustion. The moon doesn't have an oxygen-rich atmosphere, so your engine can't burn fuel to generate power. Additionally, your rubber tires would crack or melt on the surface, where temperatures range from that of liquid nitrogen to boiling water.

The upgrades are fairly straightforward. You could swap your Firestones for a set of NASA's metal-mesh, lunar-grade tires. You'd need to get rid of that combustion engine, too. An electric engine running on hydrogen fuel cells would perform best in the lunar environment, Wilcox says.

METAL-MESH
LUNAR-GRADE TIRES

BE CAREFUL OUT THERE

Even in your new Moonmobile, you'd want to keep your trips brief. On the surface of the atmosphere-free moon, there's no protecting yourself from cosmic rays, which at lunar-intensity levels can increase your risk of developing cancer by 3 percent in just six months. If you're the cautious type, you might install water-filled panels that are at least 2 inches (5 cm) thick to block the protons spewed from the occasional solar flare, which could kill you in less than an hour.

Once you'd made these modifications, you would reap some nice benefits. Because gravity on the moon is one-sixth that of Earth, your engine wouldn't have to work as hard to propel your car, so you'd score around six times as many miles per battery charge as you would here. And there's never any traffic. Of course, you'd have to get your car up there. NASA's going freight rate to the moon runs around $25,000 a pound, so delivering a one-ton car would cost $50 million. Suddenly, those two-seater buggies left behind from the Apollo missions don't sound so bad after all.

ELECTRIC ENGINE

WATER-FILLED PANELS

005
DOES THE SPACE SHUTTLE'S COMPUTER
REALLY RUN ON JUST ONE MEGABYTE OF RAM?

SO LAST MILLENNIUM

It's true: The brain of NASA's primary space vehicle has the computational power of an IBM 5150, that clunky '80s icon that goes for $20 at yard sales. According to NASA and IBM, the shuttle's General Purpose Computer (GPC)—which controls, among other things, the entire launch sequence—is actually an upgrade of the 500-kilobyte computer that the shuttle flew with until 1991.

KEEPING IT SIMPLE

Such an antiquated computer works just fine for NASA. The shuttle doesn't need to run a powerful graphics engine, create PowerPoint presentations, or store MP3s. It focuses entirely on raw functions—thrusters on, thrusters off—which, though mathematically complex, don't require the juice that a user interface like Windows calls for. The GPC has flown so many missions with hardly a hiccup that there's no reason to replace it, even if it is just 0.005 percent as powerful as an Xbox 360. Besides, doing a complete overhaul would be horrendously expensive. The GPC's software would have to be completely reconfigured for a modern computer and tested until proven flawless.

For proof that you shouldn't fix a space computer if it ain't broke, consider Russia's *Soyuz* space capsule, which since 1974 has been running Argon-16 flight-computer software with just six kilobytes of RAM. In 2003, the Russians rewrote some of the spacecraft's software, which experts suspect led to its subsequent crash-landing in a desert in Kazakhstan.

▶ **A BUMPY RIDE**

Despite its gusty reputation as a "gas giant," Jupiter's blood-red clouds hide a dense, rocky core that's perhaps 20 times as massive as Earth. That core blocks any spacecraft's passage through the center of the planet, but even a detour through the clouds would be a disaster.

Knowledge of Jupiter's innards is scarce, mostly coming from the Galileo probe, which in 1995 plunged 100 miles (161 km) into the Jovian atmosphere and relayed data until it vaporized an hour later. But here's what we do know: First, any spacecraft would need to navigate through Jupiter's instrument-scrambling radiation belts, the harshest of which extend 200,000 miles (321,869 km) from the planet. Then it would face winds of up to 230 miles per hour (370 kph) tearing across the surface of the planet's turbulent hydrogen-cloud atmosphere, and if it made it past those, gusts of nearly 400 mph (644 kph) would soon kick in. In the outer atmosphere, average temperatures run to around 306°F (152°C), and scientists suspect that it's up to 50,000°F (27,760°C) closer to the core. The atmosphere likely ripped apart the 1.2-mile-wide (1.9-km) Shoemaker-Levy 9 comet when it hit Jupiter back in 1994. Just saying.

Some 9,000 miles (14,484 km) farther in, sandwiched between the atmosphere and the hot, rocky core, Jupiter's interior most likely consists of liquid metallic hydrogen. The highly conductive fluid can exist only under space shuttle–crushing conditions like the planet's 44 million pounds per square inch (3 million kg/cm²) of pressure. So with all this in mind, although it may mean a longer trip to Saturn, we suggest taking a left turn around Jupiter.

007
COULD
ROBOT ALIENS EXIST?

▶ **TAKE ME TO YOUR PROGRAMMER**

Steven Dick, NASA's former chief historian and an astrophysicist specializing in astrobiology and the postbiological universe, thinks robot aliens could easily exist outside of B movies.

"The existence of a race of sentient alien robots might be not just possible, but inevitable. In fact, we might be living in a postbiological universe right now, in which intelligent extraterrestrials somewhere have exchanged organic brains for artificial ones.

"The driving factor is a pragmatic desire to improve mental capacity. Alien beings may have already reached a point in their evolution where, having exhausted the potential of their biological brains, they have taken the next logical step and opted for robotic brains equipped with artificial intelligence.

"This brain swap may not be as far off for humans as one might think. In only a few decades, the computer revolution here on Earth has produced supercomputers capable of performing more than a quadrillion calculations per second. (According to research by Hans Moravec, an artificial intelligence expert at Carnegie Mellon University, that rate trumps the human brain's estimated top speed of 100 trillion calculations per second.) Some scientists speculate that in a few decades, an event called the technological singularity will occur, and machines armed with computer brains will become sentient and surpass human intelligence. Civilizations equipped with technology light-years ahead of our own could have already experienced the singularity thousands, or even millions, of years ago.

"How likely is it that such a robotic race exists? Given the limitations of biology as we know it, the force of cultural evolution, and the imperative to improve intelligence, I'd say the chances are more than 50/50. That said, if postbiological beings do exist, they might not be interested in us. The gulf between their minds and ours might be so great that communication is impossible, or they might consider meatheads like us too primitive to warrant their attention."

008
CAN A LASER
CUT THROUGH
A MIRROR?

▶ **REFLECT THIS**

Sorry, Buck Rogers, but you're going to need more than a pocket mirror to fend off robot aliens armed with laser guns. Although on TV a laser bounces off a mirror, in reality that beam could easily burn right through it.

SEVEN YEARS' BAD LUCK

No mirror can reflect all the light shone at it. Some fraction of it passes through the glass coating and is absorbed by the mirror's reflective backing, typically aluminum or silver. Concentrate a powerful laser on this spot long enough, and the metal backing will absorb the laser radiation and heat up, and the beam could eventually pierce the mirror. (It's worth noting that the mirrors used to focus lasers are specially designed to absorb less light. Which really doesn't help you against those robot aliens.) In fact, pulse lasers, which are fired in concentrated bursts instead of a continuous stream, are routinely used to cut glass or mirrors. By narrowly focusing these high-energy blasts, the light can knock away the electrons in chemical bonds, causing the material to break away. You might have even seen this up close and very, very personal: The technique is often used to slice below the surface of the eye during corrective vision surgery.

009

DOES NASA HAVE A RESCUE PROTOCOL IF AN ASTRONAUT FLOATS AWAY DURING A SPACE WALK?

THE TIES THAT BIND

It's never happened, and NASA feels confident that it never will. For one thing, astronauts generally don't float free in space. Outside the International Space Station, they're always attached to the station with a braided-steel tether that has a tensile strength of 1,100 pounds (499 kg). If it's a two-person space walk, oftentimes the astronauts are also hooked to each other.

Should the tethers somehow fail, however, astronauts have an awesome backup plan: jetpacks! Each one wears what's called a SAFER, for "Simplified Aid for EVA (extravehicular activity) Rescue," a backpack with a small built-in nitrogen-jet system that can propel him or her back to the station.

Of course, SAFER is useful only if the astronaut is conscious. What if an astronaut gets bonked on the head, becomes untethered, and can't operate the jetpack? "A rescue effort could and would be undertaken by the second space-walker or other members of the space-station crew," says Michael Curie, a spokesman for NASA's space operations. He wouldn't speculate on the exact steps a rescue team would take, because they would depend on the circumstances. But he adds, "We are really happy with the tether-and-SAFER approach."

Jim Oberg, a space journalist who worked at the space shuttle's mission-control center for 22 years and specialized in rendezvous procedures, weighs in on the options for rescue. The station's robotic arm, he explains, is usually not within range of where the astronauts work and moves too slowly to grab someone. The Soyuz vehicles need a full day to power up and undock. By then, the carbon dioxide filters in the astronaut's space suit would run out, asphyxiating him. And the ISS cannot redirect its positioning rocket quickly enough to catch up to a runaway astronaut.

In a worst-case situation, the only rescue option, according to Oberg, would be for a second astronaut to link together several tethers end-to-end, attach them to the station, and then use his SAFER pack to jet over to his crewmate and haul him in. Certain conditions could make a rescue easier, he says. If an astronaut floated away more or less at a right angle from the station's orbit, orbital dynamics (which require too much math to explain here) dictate that he would float back toward the station in about an hour.

()1()
IF AN ASTRONAUT DIES ON MARS, WHAT SHOULD BE DONE WITH THE BODY?

> **WORST-CASE SCENARIO**
> Although we haven't run into this particular problem yet, Paul Root Wolpe, a bioethics adviser for NASA and the director of the Center for Ethics at Emory University, might have to grapple with something similar one day. Keep in mind that his opinion on future scenarios is not the official stance of the space agency.

"I can say with fair confidence that if an astronaut died on a short mission to the moon, the craft would turn around and come back. But it gets thornier if the astronauts are on Mars, or even halfway there—any place where turning back would be inadvisable or even impossible.

"There are really only two options for the body: Leave it there or bring it home. My guess is that NASA would make every reasonable effort to bring the body home. Returning the body would most likely be incredibly important for the other crew members, who would have formed an extremely strong bond with one another during the three-year mission. Although the astronauts chosen for this mission would have such a demeanor that they would be less likely to freak out about sharing the ride home with a dead body, they may need to undergo grief counseling en route. In addition, when a person dies, his or her body becomes the legal property of the next of kin, who would likely want to have the body returned. NASA would certainly take such a request into consideration.

"The cause of death could be a huge factor in the decision. If the astronaut died by falling into a canyon, retrieving the body could put other crew members at risk. There's also the extremely remote chance that the astronaut's suit could suffer a breach, and he or she could become infected with a deadly organism that could endanger the rest of the crew—and Earth. There is no evidence that any such dangerous organism, or any organism at all, exists on Mars, but there still needs to be a plan in place for this scenario. Without a way to contain its spread, we'd have to leave the body behind. But this, in turn, raises concerns about contaminating Mars."

011
COULD AN ASTEROID IMPACT KNOCK THE MOON INTO THE EARTH?

▶ ORBITAL BILLIARDS

"If an asteroid hits the moon, it will just get another crater," says Gareth Wynn-Williams, an astronomer at the University of Hawaii. It would take a moon-sized object to move the moon, says Clark Chapman, a planetary scientist at the Southwest Research Institute, and most likely the moon wouldn't survive. Hitting it with a much larger, denser object would be like whacking an egg with a golf club.

Let's say, though, that the moon and the thing hitting it would react like solid billiard balls. None of the known asteroids larger than 60 miles (96.6 km) in diameter orbit anywhere near the moon. But what if the largest known asteroid, Ceres—which at 600 miles (965.6 km) across is roughly the size of California and Nevada combined—did manage to slip out of its place in the asteroid belt and set out on a collision course for the moon? Hardly a budge, Wynn-Williams says. It's the equivalent of a four-year-old trying to knock over an NFL lineman. The moon orbits the Earth at some 0.6 miles (1 km) per second. This orbital momentum is so great that it would overwhelm the impact force of a collision and just continue zinging around the planet.

012
OK, SO AN ASTEROID CAN'T DO IT. WHAT WOULD IT TAKE?

▶ RISING TIDE

At minimum, you'd need an object of the same size and density as the moon to hit it at the same speed and in the opposite direction of its orbit. This could stop the moon in its tracks, and it would fall onto Earth. Even if the collision only pushed the moon into a lower or less-circular orbit, that doesn't mean we would escape unscathed: If the moon's new orbit halved its current distance from the Earth, ocean tides would get about eight times as big, Gareth Wynn-Williams says. "A lot of New Yorkers would get very wet."

COULD A MOON BASE SPEED UP THE MOON'S ORBITAL DECAY? 013

A NEW SPIN

Scientists have determined that the moon creeps 1.5 inches (3.8 cm) farther away from Earth every year, slowing the Earth's rotation. But could adding mass to the moon's surface alter its orbit even more, just as a blob of mud affects a baseball's spin?

NASA's on-again, off-again desire to build its first lunar base is ambitious, but it won't move the moon. The moon's mass is about 80 billion billion tons (7 x 10^{22} kg), which easily dwarfs the few dozen tons of construction that NASA has proposed. To put this in perspective, imagine a flea sitting on the Great Pyramid and trying to nudge the giant tomb a few inches to the left.

GRADUAL SEPARATION

Unlikely as it is, though, adding mass to the moon could theoretically change its orbit. Scott A. Hughes, an associate professor of astrophysics at the Massachusetts Institute of Technology, calculates that we would need to send more than 540 trillion tons (490 quadrillion kg) to the moon to push it an additional 1.5 inches (3.8 cm) away. That, however, would require well over a billion flights of any of NASA's existing or proposed transport vehicles.

The Earth-moon separation will ultimately lengthen our days and severely diminish the magnitude of high and low tides, says Jim Bell, a professor of astronomy at Cornell University. But this won't happen for a few billion years. And by then, the sun will be burning away the Earth's atmosphere and oceans, so we'll have more pressing things to worry about.

014

WHY NOT JUST **DISPOSE** OF **NUCLEAR WASTE** IN THE **SUN?**

▶ **A STELLAR INCINERATOR**

On paper, this is a fantastic way to wipe our hands clean of all that pesky waste. The sun is a constant nuclear reaction that's about 330,000 times as massive as Earth; it could swallow the tens of thousands of tons of spent nuclear rods as easily as a forest fire consumes a drop of gasoline. And NASA currently has two probes orbiting the sun, so the technology exists to get the job done. Alas, the benefits fall far short of the risk involved.

There isn't a space agency or private firm on the planet with a spotless launch record. And we're not talking about cheapo rockets—in March 2011, a relatively simple craft carrying NASA's $424-million Glory satellite fizzled out and crashed into the southern Pacific ocean. It's a bummer when a satellite ends up underwater, but it's an entirely different story if that rocket is packing a few hundred pounds of uranium. And if the uranium caught fire, it could stay airborne and circulate for months, dusting the globe with radioactive ash. Still seem like a good idea?

015

BRING YOUR SHADES

Of all the bodies in our solar system, the sun is the one we want to give the widest berth. It gushes radiation, and even though its surface is the coolest part of the star, it still burns at about 9,940°F (5,504°C), hot enough to incinerate just about any material. As such, there are no plans to send a manned mission in its direction any time soon (Mars is much more interesting anyway).

But it can't hurt to figure out at what distance a person would want to turn back. You can get surprisingly close. The sun is about 93 million miles (150 million km) away from Earth, and you could get about 95 percent of the way there before burning up.

That said, an astronaut so close to the sun is way, way out of position. "The technology in our current space suits really isn't designed to withstand the rigors of deep space," says Ralph McNutt, an engineer working on the heat shielding for NASA's Messenger probe. The standard space suit will keep an astronaut relatively comfortable at external temperatures reaching up to 248°F (120°C). Above that point, the suit would transform into a close-fitting sauna—the temperature inside would climb above 125°F (52°C), and the person would become dehydrated and pass out, eventually dying of heatstroke.

CATCH A LIFT

Now, the space shuttle has no business heading toward the sun either, but riding in such a vessel, someone could get much closer. The ship's reinforced carbon-carbon heat shield is designed to withstand temperatures of up to 4,700°F (2,593°C) to ensure that the spacecraft and its passengers can survive the friction heat generated when it reenters the atmosphere from orbit. If the shield wrapped the entire shuttle, McNutt says, astronauts could fly to within 1.3 million miles (2.1 million km) of the sun (roughly 98 percent of the way there). But the integrity of the shield degrades rapidly above 4,700°F (2,593°C), and the cockpit would begin to cook. "I would advise turning away from the sun well before that point," McNutt says. Much hotter than that, the shields would fail altogether, and the vehicle would combust in less than a minute.

Of course, just getting that close to the sun would be quite an accomplishment, says NASA radiation-health officer Eddie Semones. The constant exposure to cosmic radiation during the voyage would most likely prove fatal before the astronauts even reached the halfway point.

016
SAY I'M IN THE LARGE HADRON COLLIDER, AND IT'S REVVING UP. SHOULD I BE CONCERNED?

▶ **WATCH OUT FOR QUARKS**

Well, it's never a great idea to stand next to a machine that could (theoretically) create a black hole, but the magnets that steer the proton beams around the planet's most powerful particle accelerator would probably spare you from excess radiation. Then again, there is the off chance that some 300 trillion protons could erupt from the device and kill you on the spot.

Even though the LHC's twin beams travel in protective isolation through pipes 17 miles (27.3 km) long and 2 inches (5 cm) wide sucked to a near-perfect vacuum, some of those protons—potentially billions—inevitably wander off the track. When they do, they slam into the magnets that steer and focus the beam or hit other hardware, gas molecules, or protons. These collisions generate a mess of secondary radioactive particles, explains Mike Lamont, an LHC machine coordinator, filling the tunnel with a field of radiation roughly equivalent to that of a full-body CT scan. That's not a dangerous amount of radiation to be exposed to for a few minutes, but if you're planning to hang out around the LHC for a day, you very well might suffer some cellular damage.

017
WHAT IF ENGINEERS LOSE CONTROL WHILE I'M IN THERE?

▶ **RUNAWAY PROTONS**

We should make one thing clear: The LHC's highly strict security measures mean that it's virtually impossible to sneak into the tunnel when the beam is on. But let's say that you have crazy ninja skills and manage to sneak in there. Bad news. If engineers were to lose control of the beam, you're toast. The beam is only 1 millimeter wide, yet it contains 320 trillion protons moving just shy of the speed of light. That gives it about the same momentum as a train weighing 400 tons (362,874 kg) speeding at 95 mph (153 kph). It would plow through the magnets and unleash a fatal cascade of high-energy particles and radiation.

And that's just if you were near a runaway beam. If you actually stood in its way, it could burn a hole right into you, Mike Lamont says. "A human body wouldn't stand a chance."

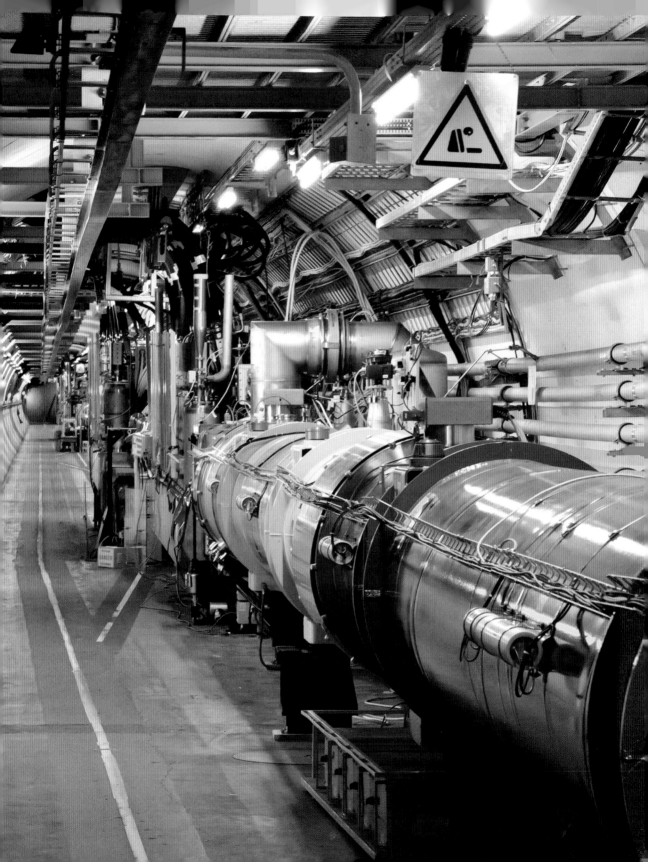

IF YOU DROPPED A CORN KERNEL FROM SPACE

WOULD IT POP DURING REENTRY?

DELICIOUS SPACE SNACKS

There's a little bit of water inside each kernel of popcorn, and if you heat the kernel above 212°F (100°C), that water should boil, turn into high-pressure steam, and pop the kernel. At least, that's what happens in your microwave oven. But in orbit, things aren't so simple. First off, the cold vacuum of space would suck all the water out of the kernel before it could pop the corn. So any ordinary kernels would be all drop, no pop.

Let's say we figured out a way to keep the kernel watertight. What then? In that case, it all depends. Anything falling through the atmosphere has what's called a terminal velocity. This is the speed at which the upward force, or drag, from air resistance equals the downward force of gravity. Typically, a falling object, like a skydiver, speeds up until it reaches terminal velocity. If something like the space shuttle starts out in the airless vacuum of space, it can reach a speed higher than its terminal velocity. But as soon as it starts passing through the atmosphere, friction will slow it down. This friction generates heat—about 3,000°F (1,650°C) for the space shuttle. Just how much heat depends on how fast the object is going, as well as its size, shape, and mass.

If an astronaut were to throw a watertight kernel out of that deorbiting space shuttle moving at 17,000 mph (27,359 kph), would the kernel reach hot enough temperatures to pop as it flew through the atmosphere? It's possible, says Kenneth Libbrecht, a physics professor at the California Institute of Technology, but he can't run the numbers to say for sure, because no one has measured how much friction a kernel generates when it moves through the air. "Of course, the other possibility is that it will heat too quickly, and the outer husk will burn off before the kernel has a chance to pop," Libbrecht says. And so, for now at least, there's no way to know. Note to the guys on the space station: please let loose a bag of Orville Redenbacher's.

HOW MUCH SPACE DEBRIS IS THERE? 019

ORBITAL TRASH HEAP

More than we'd like there to be. Debris includes spent rocket stages, pieces of broken-down satellites, and random objects, like the tools that drift off during space walks around the International Space Station. According to NASA, there are some 18,000 objects in orbit larger than a softball; 300,000 that are larger than a nickel; and millions that are even smaller (among them, the 150,000 sugar cube–sized chunks produced when the Chinese military shot down one of its defunct satellites in 2007).

This generation of trash will breed even more as the softball-sized pieces slam into and break up bigger objects, creating a still larger debris cloud. But even then, most of the new stuff will burn up on reentry, so unless you make regular trips into orbit, you can probably put space debris out of your mind.

020 WHAT'S THE CHANCE THAT FALLING SPACE DEBRIS WILL HIT ME?

THE SKY IS (ONLY OCCASIONALLY) FALLING

No need to don a hard hat. The odds that one of the millions of pieces of trash in orbit around Earth will hit you are about one in a trillion, says Bill Ailor, director of the Center for Orbital and Reentry Debris Studies.

The risk that *someone* will get hit can run far higher, though, says Nicholas Johnson, NASA's chief scientist for orbital debris. NASA and other space agencies aim to keep the risk of injury from falling objects lower than 1 in 10,000. The risks typically run higher with large objects. For example, there's a 1-in-1,000 shot that the Hubble Space Telescope could hit someone if it falls from orbit once it's decommissioned, so NASA will preemptively steer it into the ocean.

There's only one recorded instance of debris hitting a person. In 1997, a woman named Lottie Williams was in a park in Tulsa, Oklahoma, when a DVD-sized piece of metal mesh from a spent Delta II rocket hit her shoulder. It fell at what NASA calls "a very low speed," and she walked away unscathed. If you're a space tourist, however, even the tiny stuff is dangerous: The flotsam orbits at 18,000 mph (28,968 kph)—fast enough for a fleck of paint seven-thousandths of an inch (0.18 mm) wide to gash a shuttle's window in 1983.

[]21 HOW CAN I SEND SOMETHING TO SPACE ON THE CHEAP?

▶ YES, MY HAT HAS BEEN TO SPACE

For more than a decade, private companies have been happy to let you pay them to launch your personal payload into space. The good news is that the cost of room on both orbit- and suborbit-bound rockets has already plummeted, with prices now lower than $500 a pound (0.45 kg) from a number of providers, including Interorbital Systems in California, the Russian Shtil-3A rocket, and Bigelow Aerospace. The cheapest option may be from Masten Space Systems in Santa Clara, California. For a mere $99, you may soon be able to send a "SodaSat" containing any object of your choosing—maybe the ashes of a loved one, maybe some cell cultures—62 miles (100 km) into space and back as long as it weighs less than a pound (0.45 kg);

fits into a soda can–sized canister; and is not narcotic, radioactive, or explosive. The company's optional pressurized containers will even allow clients to send something that's alive, as long as it's hardy enough to survive the trip (bacteria, for example). You *do* have to be doing a scientific experiment, though, explains Masten vice president Michael Mealling: "You just can't send your cat up in space to flail around in zero G."

Masten Space Systems was formed in 2004 when David Masten cashed in his stock options from computer-network behemoth Cisco Systems to fund the start-up. The initial idea was to get the cost of flying a rocket down so low that the major expense became the fuel. Although it's still a work in progress, Masten's program is one of the best deals around.

[]22 HOW COULD SPACE FLIGHTS BE MADE SO CHEAP?

▶ ROUND-TRIP ROCKET

Masten's company is working on a line of reusable rockets. Called eXtreme Altitude, or XA 1.0, the vertical-takeoff/vertical-landing suborbital craft will land seat down, nose up, requiring just a refuel of liquid oxygen and isopropyl alcohol before launching again three hours later. Because it will land in its relaunch location and configuration, the XA 1.0 will be able to make three or four trips a day, instead of three or four trips a year, carrying up to 200 pounds (90.7 kg) of cargo at a time. An added bonus: in the middle of each 20-minute trip, the top of the rocket can open like a clamshell, exposing the loot to the vacuum of space.

023
IF WE BUILT A SPACE ELEVATOR,
WOULD IT MESS UP EARTH'S ORBIT?

▶ EH, MAYBE A LITTLE

Not enough to lose sleep over. One proposed space elevator—a permanent structure to economically transport material into space—would extend from an oceanic platform located on the equator to an 800-ton (725,747-kg) counterweight 62,000 miles (99,779 km) above the Earth's surface. Every time the elevator's "car" ascended with a payload weighing 15 tons (13,608 kg), the extended mass would increase the length of an average day by a femtosecond. (It's the same idea as when a spinning ice-skater puts his arms out to slow down.)

THERE ARE BIGGER THINGS TO WORRY ABOUT

Brad Edwards, an expert on the space-elevator concept and president of X-Tech, a developer of high-strength materials, says it's hardly something to be concerned about. Contemporary events such as melting glaciers (which move millions of tons of water from the poles to the equator) slow down the Earth's rotation orders of magnitude more than one trip on a space elevator would. Moreover, the elevator would not affect the Earth's movement around the sun in the slightest.

The only way to seriously hinder the rotation of the Earth would be to deploy about 1 percent of its mass. "You're talking about launching something that's roughly the same mass as the moon," Edwards says. "That's just not going to happen."

024

COULD WE USE THE
SOON-TO-BE-RETIRED
SPACE SHUTTLES AS SPACE STATIONS?

▶ A TRICKY RETROFIT

It seems like a good idea; after all, not many vessels are capable of sustaining life in space, so why not recycle what we've got? Unfortunately, the current fleet just isn't cut out for long-term habitation. As NASA retires the three remaining space shuttles, the craft will be sent to museums, not back into orbit.

The main problem is power. Each shuttle has three fuel cells that mix liquid hydrogen and liquid oxygen, and that reaction generates all the crew's electricity (as well as its clean drinking water). But the shuttle stocks only enough oxygen and hydrogen to keep the fuel cells running for 14 days. After that, the shuttle would go completely dark. It wouldn't be able to maneuver or send radio transmissions, and life-support systems would shut down. "You couldn't even cook breakfast," says Michael Curie, a spokesperson for NASA space operations. "You wouldn't be able to return to Earth. Basically, the shuttle would be space junk."

You could restock the fuel tanks, of course, but the shuttles themselves have served as NASA's biggest orbital delivery trucks, and no appropriate replacement is in the works, so there really isn't an easy way to refuel the "shuttle stations." The International Space Station gets around the energy issue by using its giant solar panels to charge its batteries. Outfitting the shuttles with a similar system would take quite a bit of engineering. "At that point," Curie says, "it would make more sense to start from scratch."

But what really makes a shuttle station a bad idea is the lack of amenities. The shuttles don't have room for all the exercise equipment that astronauts need to stave off rapid bone and muscle loss. They don't have individual bedroom compartments like the ISS does; to get some shut-eye, astronauts instead zip themselves in sleeping bags and Velcro themselves to a wall. They don't even have a garbage chute. "They'd have to figure out some way to bundle up waste—human waste included—and toss it out a hatch," Curie says. "And because there isn't a launcher to shoot the waste into the atmosphere to burn up, it would just float and collect around the outside of the station."

HOW LOUD IS IT INSIDE THE INTERNATIONAL SPACE STATION? 025

PACK SOME EARPLUGS

Pretty loud. Noise levels on the ISS, which has been in low-Earth orbit since 1998, are between 55 and 78 decibels—the equivalent of the range between the volume of a normal conversation and a lawn mower. The noise comes from the fans that circulate air throughout the station. In the past, the ISS averaged about 10 decibels louder.

NASA has been working to lower noise levels through various muffling methods, including fan-vibration isolation, acoustic paneling on walls, and mufflers on fan outlets, and by replacing older fans when they break down with more efficient models. NASA spokesperson Kylie Clem says that, with these repairs, ISS noise levels are well on their way to the space agency's goal of 50 to 60 decibels.

026 IS ALL THAT NOISE A PROBLEM FOR THE ASTRONAUTS?

WORKING OVERTIME

Astronauts spend as long as six months on board during research missions, so all this racket is a genuine concern. On Earth, noise levels below 85 decibels are considered acceptable in the work environment (even without ear protection) by the National Institute for Occupational Safety and Health. There is no standard for continuous exposure beyond the eight-hour workday, however. And there's no way to get away from the noise on the ISS, notes Richard Danielson, the manager for audiology and hearing conservation at the National Space Biomedical Research Institute. "The crew member who works, eats, and sleeps in the service module is there all the time without a rest." Apart from a Russian news service article mentioning temporary diminished hearing capacity, there is no hard evidence available that points to astronaut hearing loss. But a quieter ISS certainly can't hurt.

WHY DON'T MOONS HAVE MOONS?

A MOON OF ONE'S OWN

Astronomers can say with near certainty that there are no moons with moons in our solar system. But that doesn't mean it's physically impossible. After all, NASA has successfully put spacecraft into orbit around our moon.

Although astronomers have spotted some asteroids with moons, a parent planet's strong gravitational tug would make it hard for a moon to keep control of its own natural satellite, says Seth Shostak, a senior astronomer at the nonprofit Search for Extraterrestrial Intelligence (SETI) Institute. "You would need to have a wide space between the moon and planet," he says. Orbiting far from its parent planet, a relatively massive moon might be able to hold onto a moon of its own.

OUT OF SIGHT

Conditions like these might exist in far-off solar systems, but while hundreds of exoplanets (planets outside of our solar system) have been detected, there's almost no chance we'll be able to spot exomoons, much less moons of exomoons, for decades to come. Most planet-hunting methods—such as spotting one as it passes a large star—lend themselves to detecting mostly huge, Jupiter-like planets, or sometimes Earth-sized rocky planets, but not their moons.

Even if astronomers spot a moon with a moon, it probably won't last long. "Tidal forces from the parent planet will tend, over time, to destabilize the orbit of the moon's moon, eventually pulling it out of orbit," says Webster Cash, a professor at the University of Colorado's Center for Astrophysics and Space Astronomy. "A moon's moon will tend to be a short-lived phenomenon."

DECEPTIVELY QUIET

Claims that the moon was anything other than a old, dead rock used to be the domain of crackpots. As far as scientists can tell, the last volcanic eruptions on the surface of our steady companion occurred anywhere from one billion to three billion years ago, when the basaltic plains of the lunar seas were formed. Since then, the only changes to the surface are thought to have resulted from a constant bombardment of meteorites.

In 2006, however, geologist Peter Schultz of Brown University and his colleagues published a fresh examination of photos from the Apollo missions and data from satellite observations of the lunar surface. It revealed a hill on the moon that looked fresher than it should, probably no more than 10 million years old. The researchers hypothesize that occasional bursts of gas from surface fissures blow as much as 30 feet (9 m) of dust off the mile-wide formation, called Ina, hinting at ongoing internal processes not previously suspected. The eruptions, which could have occurred at any time in the past 10 million years, could be the product of a still-cooling magma core, or perhaps pockets of trapped gas rupturing from the moon's crust, but Schultz stops short of crediting volcanic activity for the burps. "These are not explosions of lava coming out, and since we

haven't actually seen [the gas bursts], we don't know if they're hot or not," he says. "But my guess is that this is very likely cold gas that may have been hot deep in the interior."

VOLCANO MISSION

Schultz's team hasn't yet determined exactly what gases are spewing from the fissures—although NASA's Lunar Prospector satellite detected radon, polonium, and carbon dioxide near the surface—but he doesn't expect the eruptions to hinder future lunar missions or bases. "In fact," Schultz says, "it might be a good place to go and explore."

WHERE'S MY HOVERBOARD?

It's almost 2015 and for a long time, there was no sign of anything like Marty McFly's supercool hoverboard from *Back to the Future II*. Until now. Nils Guadagnin, a French artist, has built a perfect replica of Marty's hoverboard from the film. Guadagnin uses electromagnets to keep the board floating, and a laser system keeps it stable. Unfortunately for us, it won't support a rider—Guadagnin considers the project a sculpture—but it keeps the hope alive that 1985's dreams of the future could still come true.

COULD *STAR TREK*-STYLE REPLICATORS EXIST?

We may not be able to make a delicious sandwich from thin air and stray molecules quite yet. But a new technology developed by NASA scientists called Electron Beam Freeform Fabrication, or EBF3, will allow users to generate any metal object from a 3D rendering. In EBF3, an electron beam inside a vacuum chamber is focused on a source of metal. The beam melts the metal according to the 3D drawings of the object being produced, ending with a finished metal part.

This technology allows for uniform production of metal parts even far away from complex, dedicated factory facilities, and uses significantly less metal than standard production procedures. And in the future, manned expeditions to Mars or the moon could use EBF3 to produce specialized equipment using minerals mined onsite. Not quite *Star Trek*, but not too shabby.

WOULD IT BE POSSIBLE TO MAKE AN INVISIBILITY CLOAK?

Harry Potter may soon lose his monopoly on invisible couture. Researchers at Lawrence Berkeley National Laboratory and Cornell University have recently designed invisibility cloaks that work in the visible-light spectrum. These nano-scale cloaks redirect light around an object, fooling an observer into perceiving a flat space instead of a telltale bulge. They're pretty tiny at the moment, though—neither would obscure much more than a poppy seed—so you may need to wait before placing an order for your dream cape.

Adding a little bling to the mix, scientists from Tufts and Boston University have created an invisibility cloak made from silk coated with gold. It doesn't work in the visible-light spectrum yet (it only functions in the terahertz range, between radio and infrared), but researchers are optimistic that the technology could be refined to work in the range of visible light. The runways will never be the same.

CAN WE CREATE HOLOGRAPHIC VIDEOS?

Need to send out an interplanetary distress signal à la Princess Leia's robot-relayed message to Obi Wan? Never fear: A team at the University of Arizona College of Optical Sciences has created a new holographic imaging technology that records and projects a three-dimensional, moving image in real-time.

The new device has a 10-inch (25.4-cm) screen that can refresh every two seconds. That's nowhere near the rate of normal film yet, but it's a big leap forward from the static images that were the extent of holographic technology until recently. And once the refresh rate speeds up, who knows? We may be chatting in 3D before we know it.

033 WOULD A SLINKY WORK ON THE MOON?

MOONWALKING

Most anyone above a certain age could tell you what walks down stairs, alone or in pairs, and makes a slinkity sound. But would Slinky walk down stairs on the moon? Tom James, son of the toy's inventor and self-proclaimed Slinky master, professes doubt. "It would just flip back and sit again," he says. "It might work on Mars, but who the heck's going to prove or disprove that?"

Enter astrophysicist Marc Kuchner of the NASA Goddard Space Flight Center, who used a standard-issue plastic version of the coiled toy to develop his "Slinky hypothesis" for how it would act on other planets. As with any spring, Kuchner explains, a Slinky is simply a bundle of stored energy, and once you supply the push to start it down the stairs, gravity and inertia take over and continue flipping the spring until it reaches the bottom step.

Because the moon's gravity is just one-sixth that of Earth's, Kuchner calculates that a lunar Slinky would walk at about 40 percent of its typical speed. The plastic variety would need steps that were at least 7 inches (17.8 cm) high, however, to sustain motion down them. (The smallest steps Kuchner's Slinky would walk down on Earth were just one inch [2.5 cm] high.)

034 WHAT OTHER FACTORS WOULD AFFECT SLINKY PERFORMANCE?

IT'S FUN FOR A GIRL AND A . . . MARTIAN?

Gravity is key: A Slinky would work much better, albeit more slowly, on Mars, where gravity 38 percent that of Earth's would allow it to crawl down stairs as short as 3 inches (7.6 cm). Materials are important, too. You'd be best off with a plastic version rather than the classic metal, Kuchner adds. A plastic Slinky requires less energy to slink than a metal version, so it would have an easier time moving in low gravity. Also, the Red Planet's super-oxidizing conditions would turn the steel to rust in about a month.

Temperature also would play a role: For example, on the day side of the moon, where sunlight cranks the heat to 260°F (127°C), the toy would slink faster than on the night side, where temperatures as cold as -280°F (-173°C) would stiffen the spring. And because the moon more or less lacks an atmosphere, which is necessary to propagate sound waves, don't expect to hear any slinkity sound at all.

035
WOULD A HELIUM BALLOON FLOAT ON THE MOON?

▶ **GROUNDED**

A helium balloon on the moon might as well be made of lead. For any balloon to stay aloft in any atmosphere, the gas inside it must be lighter than the surrounding air.

Ultralight helium has no trouble climbing in Earth's atmosphere, which consists mostly of heavy molecular nitrogen and oxygen. "But on the moon, there is no air, so there's nothing for the helium to rise above," says Marc Rayman, an engineer at NASA's Jet Propulsion Laboratory. Unable to escape even lunar gravity, which is one-sixth that of Earth's, the balloon would plunge to the ground.

Try throwing a birthday party on the International Space Station, and you'll run into the opposite conditions. Because atmospheric pressure in the station is kept the same as at sea level on Earth, a helium balloon would have the support it needs to float, but with almost no gravity on the station, there is no force to push up or down on the balloon. "It would just hang there, the same as if you let a hammer go," Rayman says.

If you're looking to throw an extraterrestrial fiesta for an astronaut, book a spot on Mars. The thin air on the Red Planet is still heavier than at the highest altitudes on Earth where balloons have floated. That, combined with gravity that's about a third of Earth's, would send a helium balloon on Mars up and away.

036
WHAT DOES
A STAR SOUND LIKE?

▶ **BONFIRE ON MUTE**

Observing a star up close (putting aside for a moment how you'd get there or withstand its heat) is probably like sitting beside an enormous silent fire. Sounds—which are simply pressure variations in a medium such as air or water—can't propagate in the vacuum of space, so the roiling surface of a star would make an impression on the eyes, but not the ears.

THIS ONE GOES UP TO 11

A supernova, however, just might have the loudest soundtrack in the universe. When a star explodes, the massive detonation expels stellar material far into space, and that matter could theoretically provide a medium through which sound vibrations might travel. Assuming you somehow survived the blast—the initial shock wave would travel up to 20,000 miles per second (32,187 km/s) and carry 1,044 joules of energy—it would sound like "10 octillion two-megaton thermonuclear devices detonated simultaneously," says Charles Liu, an astrophysicist at the City University of New York College of Staten Island. "When those stellar guts hit your eardrums, you'll hear it. That is, as long as your eardrums stay attached."

037
WHAT DOES SPACE SMELL LIKE?

▶ **HYDROCARBON PERFUME**

The final frontier smells a lot like a NASCAR race—a bouquet of hot metal, diesel fumes, and barbecue. The source? Dying stars.

The by-products of all this combustion are smelly compounds called polycyclic aromatic hydrocarbons. These molecules "seem to be all over the universe," says Louis Allamandola, the founder and director of the Astrophysics and Astrochemistry Laboratory at NASA Ames Research Center. "And they float around forever," appearing in comets, meteors, and space dust. These hydrocarbons have even been short-listed for the basis of the earliest forms of life on Earth. Not surprisingly, polycyclic aromatic hydrocarbons can be found in coal, oil, and even food.

Though a pure, unadulterated whiff of outer space is impossible for humans (it's a vacuum after all; we would die if we tried), we can get an indirect sense of the scent: When astronauts are outside the International Space Station, spaceborne compounds adhere to their suits and hitch a ride back into the station. Astronauts have reported smelling "burned" or "fried" steak after a space walk, and they aren't just dreaming of a home-cooked meal.

The smell of space is so memorable and distinct that, three years ago, NASA reached out to Steven Pearce of the fragrance maker Omega Ingredients to re-create the odor for use in its training simulations. "Recently we did the smell of the moon," Pearce says. "Astronauts compared it to spent gunpowder."

Allamandola explains that our solar system is particularly pungent because it is rich in carbon and low in oxygen, and "just like a car, if you starve it of oxygen you start to see black soot and get a foul smell." Oxygen-rich stars, however, have aromas reminiscent of a charcoal grill. Once you leave our galaxy, the smells could get really interesting. In dark pockets of the universe, molecular clouds full of tiny dust particles may host a veritable smorgasbord of odors, from wafts of sweet sugar to the rotten-egg stench of sulfur.

038
HOW BIG CAN A PLANET BE?

▶ **SHORT ANSWER? REALLY BIG**

It's tough to gauge "big"—are we talking largest diameter or most massive? Let's start with massiveness. The heaviest planets reach roughly 13 times the mass of Jupiter, the most massive planet in our own solar system (Jupiter's mass is about 318 times that of Earth). In an orb much more massive than that, the inside grows dense and hot enough to fuse hydrogen nuclei. That makes the object a brown dwarf, which burns far less brightly than a star yet emanates a distinctly un-planetlike glow.

But also consider the physical expanse. If you make a planet too heavy, it collapses under its own weight and gets smaller, according to astronomer Dimitar Sasselov of the Harvard-Smithsonian Center for Astrophysics. A planet with a mass greater than 13 Jupiters will actually end up being smaller in diameter than Jupiter itself.

"The sweet spot is around two Jupiter masses," Sasselov says of the point at which planets reach their maximum girth before the added mass forces them to compress. If a planet this size is made primarily of hydrogen and helium—light elements that float far apart from each other—it will achieve the maximum possible wingspan: 106,627 miles (171,600 km), or roughly 20 percent wider than Jupiter. Some planets, though, seem to defy these rules. Mysterious "puffy planets" and young planets beat the diameter limit but only because heat (in the case of young planets, from their formation) makes them expand. Sasselov has seen a planet that's nearly 40 percent wider than Jupiter. But he notes that once planets age and cool, they shrink to a more reasonable size. If only beer guts worked that way.

039

COULD THERE BE A
PLANET HIDDEN
ON THE OPPOSITE SIDE OF OUR SUN?

▶ **NO SUBSTITUTE FOR PLUTO**

The sun might seem like a pretty huge galactic blind spot, but we've already managed to glimpse behind it, and there's nothing there in the way of another Earth, says NASA scientist Michael Kaiser, "unless it's awfully tiny."

Kaiser is the project scientist for NASA's Solar Terrestrial Relations Observatory (STEREO) mission, which in 2006 sent two golf cart–sized satellites into orbit around the sun to study the explosions on the solar surface that are a major factor in space weather. A few months after their launch, the two probes flanked the sun and were angled such that they could see beyond it, but they found no planets lurking behind the big star.

Even if we couldn't see behind the sun, the gravitational pull of a planet roughly 100 miles (161 km) wide hiding there would noticeably affect the orbits of the other planets. And if astronomers had somehow missed that detail, Kaiser says, an unaccounted-for tug of gravity in the solar system would have disrupted the orbit of man-made satellites circling the Earth or thrown an intra–solar system spacecraft off course. That hasn't happened, so unless beings on a hidden planet have invented both an invisibility cloak and a gravity-masking device, the other side of the sun is almost certainly just empty space.

040
CAN GUNS FIRE IN SPACE?

HAPPINESS IS A WARM GUN

A fire needs oxygen to burn, so if you're in the atmosphereless void of space, there's no way to ignite gunpowder to fire a bullet, right? Wrong. Your Smith & Wesson should work just fine at the O.K. Space Corral.

Just ask Peter Schultz, an astronomer and planetary geologist at Brown University, who has run experiments on firing projectiles in space-like conditions. The explosion that fires a bullet does require oxygen for combustion, Schultz says, but it doesn't draw it from the air. Rather, oxygen comes primarily from an ingredient in the gunpowder. The spark created when the gun's hammer strikes the cartridge ignites the propellant—typically a nitrate-containing chemical, such as nitroglycerin or nitrocellulose—and converts the oxygen in it to its gaseous state. This rapid gas expansion generates a shock wave that propels the bullet out of the gun at high speed, says Schultz.

Before you fire, hold on to something solid. Thanks to Newton's Third Law—that's the one about reciprocal actions—the kickback after you fire a round would send your body flying backward in the near zero-gravity vacuum of space. "A person floating in space would move backwards, but their high mass would preclude moving backwards at the same speed as the bullet," Schultz says.

041
WHAT IS THE MOST POWERUL LASER IN THE WORLD?

SIZE MATTERS

When it reached full operation in late 2010, the National Ignition Facility (NIF) beamline handily put all other lab lasers to shame. The NIF system is 60 times as energetic as Nova, NIF's predecessor at Lawrence Livermore National Laboratory and the previous record holder, which generated 16 trillion watts. But achieving such intensity hasn't come easy. The hardware and electronics that power the NIF laser require a space bigger than a football stadium.

Now, for the numbers that make the hearts of laser nerds sing. On final approach, 192 laser beams zing into a million-pound (453,592-kg) target chamber 33 feet (10 m) in diameter, with walls 20 inches (50.8 cm) thick. To prevent any radiation from escaping, the chamber is encased by walls that are 6 feet (1.8 m) deep. Each 20-nanosecond laser burst will blast target materials in the chamber with 500 trillion watts of power—1,000 times the electrical output of the entire United States over the same period of time. All that awesome power will be used in fusion research, in astrophysics (for example, studying what conditions might be like at the center of Jupiter), and in generating thermonuclear detonations for weapons research.

LOCK AND LOAD

In more good news, your shot would be as true in space as it is on terra firma. The copper that encases some bullets melts as it zips down the barrel of the gun, which is lined with spiraled grooves called rifling. Because the exterior of the bullet melts, the rifling is able to spin the bullet, causing it to travel faster and on a more stable trajectory. Contrary to what you might expect, this still works in a cold, space-like vacuum. "In fact, we perform experiments in a vacuum with a rifled barrel," Schultz says. "It's not a problem."

HAVE MODERN SCIENTISTS MADE ANY ADVANCEMENTS IN ALCHEMY? []42

THE PHILOSOPHER'S LASER

Ancient alchemists were willing to try just about anything to turn lead into gold. That achievement is still beyond today's scientists, but if the alchemists had just had a laser, they might have been able to transform a boring metal into pseudo-platinum.

More than half of the world's platinum goes into the catalytic converters of automobiles, where it helps neutralize the carbon monoxide and nitrous oxide in exhaust. That demand is what drives platinum's price up. Recently, Penn State University chemist A. Welford Castleman Jr. used a laser to kick an electron off a molecule of tungsten carbide, giving it properties that allow it to act as platinum does. Tungsten carbide, a metallic compound just one-thousandth the cost of

platinum, won't pass as an engagement ring, but it could help bring down platinum's price by stepping in for the grittier jobs.

Castleman hopes to repeat the trick with groups of tungsten carbide molecules, while also finding stand-ins for other rare elements. Like the alchemists, he still hasn't figured out how to transmute lead to gold, but he's happy to mimic the rest of the periodic table.

043
HOW LONG WOULD IT TAKE
TO WALK A LIGHT-YEAR?

▶ **MAKE SURE YOU STRETCH FIRST**
If you started just before the first dinosaurs appeared, you'd probably be finishing your hike just about now.

Here's how it breaks down: One light-year—the distance light travels in one year, used as the yardstick for interstellar distances—is about 5.9 trillion miles (9.5 trillion km). If you hoofed it at a moderate pace of 20 minutes a mile (1.6 km), it would take you 225 million years to complete your journey (not including stops for meals or the restroom). Even if you hitched a ride on NASA's Mach 9.8 X-43A hypersonic scramjet, the fastest aircraft in the world, it would take more than 90,000 years to cover the distance.

BE PREPARED
You'd need to bring a big backpack, too; walking such a distance requires substantial supplies. The average adult burns about 80 calories per mile (1.6 km) walked, so you'd need in the neighborhood of six trillion granola bars to fuel your trip. You'd also produce a heap of worn-out shoes. The typical pair of sneakers will last you 500 miles (805 km), so you'd burn through some 11.8 billion pairs of shoes. And all that effort wouldn't get you anywhere, astronomically speaking: The closest star to the sun, Proxima Centauri, is 4.22 light-years away.

▶ GOING DOWN

Just getting to the center of the Earth and surviving is impossible. The Earth's core is about 9,000°F (4,982°C)—as hot as the sun's surface—and would instantly roast anyone who found himself there. Then there's the pressure, which can reach roughly three million times that on the Earth's surface and would crush you. But let's not sweat the details. Once you arrive in the center of the Earth, the physics gets really interesting.

Understanding gravity, the force of attraction between objects, is going to be key to wrapping your head around what is about to be a bizarre situation. The strength of gravitational attraction is determined by an object's mass and how close it is to another (more mass and closer together means increased force). The only gravity strong enough for us to feel comes from the Earth's mass, which is why we feel a downward pull on the surface.

At the center of the Earth, the situation is different. Because Earth is nearly spherical, the gravitational forces from all the surrounding mass counteract one another. In the center, "you have equal pulls from all directions," says Geza Gyuk, the director of astronomy at the Adler Planetarium in Chicago. "You'd be weightless," free-floating.

But what would happen if you tried to get out of the center by, say, climbing up a very long ladder that ends in Los Angeles? (For clarity's sake, let's assume that the Earth is uniformly dense. It isn't, but the general trend described here still holds.) At the center of the planet, the gravity from the mass beneath your feet all the way to the other side of the Earth, the Indian Ocean, will be "pulling" you down, even as the mass above your head is "pulling" you up, toward L.A. After climbing a few rungs, the total pull you feel down to the Indian Ocean will still be nearly zero. You will still feel almost weightless. But as you climb, there will be less and less mass above, and more and more below. The pull toward the core will feel greater and greater, and you will feel less and less weightless, until you are standing on the Earth's surface, staring up at the Hollywood sign, feeling heavy again.

045

WHY ISN'T THE
HUBBLE SPACE TELESCOPE
JUST ATTACHED TO THE
INTERNATIONAL SPACE STATION?

▶ **IMPERFECT MATCH**

It's a logical question. After all, it would be handy if every time the Hubble went on the fritz, an astronaut could reach out the window and give it a whack. Unfortunately, not only is that setup nearly impossible, but being docked to the ISS would impair Hubble's performance.

Hubble's orbit is 350 miles (563 km) above Earth and set at a 28.5-degree angle relative to the equator. Just bringing the scope to the ISS's 52-degree orientation would require a tremendous amount of rocket power, but getting it to the station's orbit 150 miles (241 km) below would kill it. The descent would generate enough atmospheric drag to damage the scope's solar panels. Worse, the space around the ISS is full of gases, liquids, and other debris jettisoned from the station that could gum up Hubble's optics.

Even if the scope could survive the trip, attaching it to the station would make it almost unusable, says chief Hubble engineer John Grunsfeld. The telescope captures such highly detailed images because it's free from any disturbances, atmospheric or otherwise. In other words, it's designed to stay very, very still. "Once its camera locks onto an object, it's unflinchable," Grunsfeld says. The vibrations from gear on the ISS would make such precise observations impossible. For now, Hubble will stay right where it is. As Space Telescope Science Institute news director Ray Villard puts it, "They just weren't made for each other."

046
COULD THE
HUBBLE SPACE TELESCOPE
PHOTOGRAPH NEIL ARMSTRONG'S
LUNAR FOOTPRINTS?

▶ JUST USE THE ZOOM LENS

Snug in Earth's orbit, Hubble is free from the background glare that earthly telescopes must fight to see the stars. This allows its supersensitive camera to take better photos of galaxies farther away—and thus much dimmer—than any optical telescope on the ground can. But despite being closer to the moon than any other telescope, there's no way the scope could snap a photo of that "one small step" man took in 1969.

Considering the distance to the moon and the resolving power of Hubble's 8-foot-wide (2.4-m) main mirror, one pixel in the highest-resolution image that the scope could take of the moon would be about the size of a football field, says Hubble astronomer Frank Summers of the Space Telescope Science Institute.

To get the resolution to the point where one pixel was the size of a footprint, he says, Hubble would need a primary mirror roughly 2,400 feet (732 m) in diameter. And if you wanted to make out the tread marks from Armstrong's boots, that mirror would need to be 9 miles (14.4 km) wide. "The mirror sizes required are absurd," Summers acknowledges.

But he does suggest a potential solution. Precisely positioning several spacecraft several miles apart and training them at the same target can approximate, in effect, a miles-wide telescope. Using a computer to combine observations from each scope might reveal footprints around the abandoned lunar rover, Summers says, but NASA's time would be better spent exploring the universe, not looking at dusty landmarks.

047 COULD ALIENS TRACE OUR TV SIGNALS BACK TO EARTH?

MUST-SEE TV

The French think so. Recently that country's National Center for Space Studies teamed up with the ARTE Channel (a Franco-German TV network) to broadcast a program called *Cosmic Connexion* on French TV—and into space. They beamed the signals toward Errai, a star system that lies 45 light-years away and warms a Jupiter-sized planet. The first-ever show targeted specifically at extraterrestrials features a man and a woman wearing only a layer of white paint and guiding potential viewers through a three-hour presentation of all things earthly, such as music videos and cartoons.

No one can say how extraterrestrials would interpret the nude twosome—the couple is an allusion to the naked man and woman depicted on a plaque affixed to NASA's *Pioneer* spacecraft in 1972—or whether they would be able to view the transmission at all. But assuming that intelligent beings in the Errai system have developed technology similar to ours by the time the signal arrives there in 2051, it wouldn't be difficult for them to locate the origin of the strange broadcast.

048 COULD ALIENS TELL OUR TRANSMISSIONS APART FROM BACKGROUND RADIATION?

LONG-DISTANCE CALL

Even after traveling 45 light-years, the television broadcast's radio signals would still be quite strong compared to the natural radio waves bouncing around the cosmos, says Frank Drake, founder of SETI. "If [the extraterrestrials] have an appropriate radio-telescope system, they can pinpoint the origin of the signal to extremely high precision." Alien astronomers could also potentially study the Doppler shift—the change in frequency as transmitter and receiver move relative to each other—to learn, for example, the size of Earth or the length of our day.

Hopefully, the folks in the Errai system are already hard at work building sensitive radio telescopes, because "if they don't know how to do that," Drake says, "then we're not going to find them and they're not going to find us."

WOULD IT BE POSSIBLE
TO BLOW UP MARS?

▶ **THE NUCLEAR OPTION**

It would be impossible to destroy the Red Planet with any device scientists can currently build, let alone finance. Planets can survive enormous assaults; the Hellas Basin, a Martian crater about 1,300 miles (2,092 km) wide, testifies to the planet having once collided with an asteroid so massive that the impact generated well over a hundred million megatons of energy. If a meteoroid that size were to hit Earth, it could wipe out life on an entire continent in a flash.

In contrast, the most powerful nuclear weapon ever tested, Russia's "Tsar Bomba," had a yield of only 50 megatons, and most countries' nuclear arsenals consist of bombs in the range of 200 kilotons to 400 kilotons—in planetary-impact terms, the equivalent of party poppers. Faced with an object as robust as a planet, there is no way a nuke—or all the existing nukes combined—would work. And even the mightiest of meteorite impacts haven't destroyed Mars or the Earth, explains planetary scientist Edward Scott of the University of Hawaii. "The amount of energy needed is so preposterous that it could never happen."

But what if we could build a radically more powerful weapon, one that would unleash, say, a billion billion megatons—roughly the amount of energy the sun produces in a month? According to planetologist Gary Peterson of San Diego State University, Mars's mass creates a gravitational field strong enough to render even this colossal effort a failure. "You could have the biggest explosion possible, one that would tear the planet apart, but the pieces of rock would just clump right back together again," he says. A more realistic and productive endeavor for nukes in space, he adds, would be finding a way to pulverize smaller asteroids flying too close to Earth for comfort.

WHAT **DON'T I KNOW**
ABOUT THE **FIRST MOON LANDING?**

▶ **LUNAR SECRETS**

The *Apollo 11* moon landing was one of the most iconic moments of the late twentieth century—an estimated 500 million people worldwide watched that "first step for man," the largest audience ever for a live television broadcast at that time. So, you'd think we'd know everything there is to know about that iconic event. And you'd be wrong. In researching his 2009 book, *Rocket Men*, journalist Craig Nelson uncovered these facts in various NASA archives.

1 The *Apollo*'s Saturn rockets were packed with enough fuel to throw shrapnel weighing 100 pounds (45.3 kg) 3 miles (4.8 km), and NASA couldn't rule out the possibility that they might explode on takeoff. NASA seated its VIP spectators 3.5 miles (5.6 km) from the launchpad.

2 The *Apollo* computers had less processing power than a modern cell phone.

3 Drinking water was a fuel-cell by-product, but *Apollo 11*'s hydrogen-gas filters didn't work, making every drink bubbly. Urinating and defecating in zero gravity, meanwhile, had not been figured out; the latter was so troublesome that at least one astronaut spent the mission on an anti-diarrhea drug to avoid it.

4 When *Apollo 11*'s lunar lander, the *Eagle*, separated from the orbiter, the cabin wasn't fully depressurized, resulting in a burst of gas equivalent to popping a champagne cork. It threw the module's landing 4 miles (6.4 km) off target.

5 Pilot Neil Armstrong nearly ran out of fuel landing the *Eagle*, and many at mission control worried he might crash. Apollo engineer Milton Silveira, however, was relieved: His tests had shown that there was a small chance the exhaust could shoot back into the rocket as it landed and ignite the remaining propellant.

6 The "one small step for man" wasn't all that small. Armstrong set the ship down so gently that its shock absorbers didn't compress. He had to hop 3.5 feet (1 m) from the *Eagle*'s ladder to the surface.

7 When Buzz Aldrin joined Armstrong on the surface, he had to make sure not to lock the *Eagle*'s door because there was no outer handle.

8 The toughest moonwalk task? Planting the flag. NASA's studies suggested that the lunar soil was soft, but Armstrong and Aldrin found the surface to be a thin wisp of dust covering hard rock. They managed to drive the flagpole a few inches into the ground and film it for broadcast, and then took care not to accidentally knock it over.

9 The flag was made by Sears, but NASA refused to acknowledge this because they didn't want "another Tang."

10 The inner bladder of the space suits—the airtight liner that keeps the astronaut's body under Earth-like pressure—and the ship's computer's ROM chips were handmade by teams of "little old ladies."

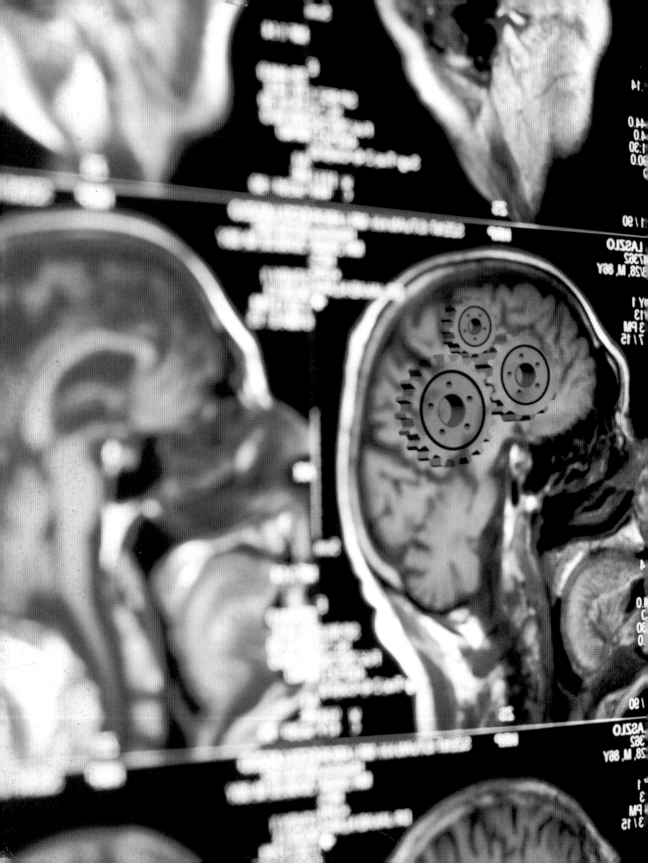

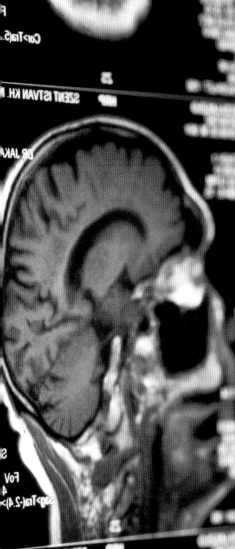

()51
IS PLAY UNIQUE TO MAMMALS?

▶ MONKEY BUSINESS

The answer appears to be no. A bird will spend hours tossing a pebble in the air, but it's nearly impossible to discern if it's goofing around or honing its talon-eye coordination. Gordon Burghardt, an animal-behavior expert at the University of Tennessee, defines play as behavior that doesn't seem to have a survival purpose, is rewarding in and of itself, and is performed when an animal is fully fed and stress-free.

PARTY IN THE JUNGLE

By that definition, the animal world is full of horseplay. Recreation is well documented in big-brained birds, such as crows and hawks, which chase each other and drop and catch objects seemingly for the hell of it. For other animals, the data is sketchier, often relying on just one case, but does suggest that play beyond birds and mammals is indeed possible. A Komodo dragon at the National Zoo in Washington, D.C., plays tug-of-war with its keepers, and soft-shelled turtles at other zoos push balls with their snouts and swim through hoops. Wild-reared octopuses entertain themselves in the lab by towing toys or passing them between tentacles. Another study documents wasps engaging in what seems to be play-fighting.

"People have observed that wasps and fish can play," says Sergio Pellis, who studies animal behavior at the University of Lethbridge in Alberta. "We're finding out that play is not just for intelligent, big-brained animals."

052
WHY AREN'T (MOST) HUMANS FURRY?

▶ **HAIRY SITUATION**

Ever since Darwin first made headlines, scientists have been pondering why humans lost their natural coats as they evolved from apes. The theories range from lice to cannibalism.

The traditional theory—refined by scientists over the past 40 years—proposes that humans gradually became furless to withstand the brutal heat of the African savannah or to prevent ourselves from overheating while chasing prey.

One alternative idea, put forth in 2003 by evolutionary biologist Mark Pagel of the University of Reading in England, is that as humans developed the ability to make clothing and shelter, they lost the extra body hair. This hairlessness prevented parasites, such as mites and ticks, from taking up residence on our bodies. Avoiding parasites led to healthier humans, Pagel posits, and because there's nothing as attractive as a bug-free hominid, hairlessness became a desirable feature in a mate, and natural selection drove the hairier folks into extinction.

In 2006, developmental psychologist Judith Rich Harris suggested a far more gruesome mechanism: As humans became hairless as a result of chance mutations, they split geographically from their hairy cousins. Once hairlessness was in style, any hirsute baby born to a hairless tribe was abandoned. As hairlessness became the norm, a thick fur coat would have become so rare that hairy humans—or near-humans, like Neanderthals—would have been seen as animals and been hunted for food. The days before waxing were savage indeed.

053
CAN THE FOOD I EAT AFFECT MY DESCENDANTS' GENES?

▶ **HOLD THE MAYO**

A recent study suggests that the same vitamins in spinach that perform instant wonders for Popeye's biceps might pack longer-lasting effects, such as dictating the hair color and health of future generations. Your lunch order could make a bigger difference than you think.

A 2006 study led by David Martin, an oncologist at the Children's Hospital Oakland Research Institute in California, tested whether a mouse's diet alone can affect its descendants. The researchers fed meals high in minerals and vitamins—such as B12, which fortifies leafy greens—to pregnant mice that have a gene that makes their fur blond and also increases the likelihood that they will grow obese and develop diabetes and cancer. On the new diet, the mice produced brown-haired offspring that were also less vulnerable to disease. Even when the second-generation mice were denied the supplements, their offspring retained the improved health and still grew dark fur coats.

Martin's study isn't the first to note this type of generation-spanning phenomenon. In 2002, Swedish researchers dug through century-old records and determined that a man's diet at the onset of puberty affected his grandson's vulnerability to diabetes. The study tracked 303 men, and those with an abundant supply of food were four times as likely to have grandchildren die of diabetes. Though far from exhaustive, the study indicated that genes are more susceptible to outside forces than has been commonly believed.

But don't start your teenager on that all-spinach diet just yet—scientists warn that the influence of diet on human gene expression is not fully understood. Nevertheless, Martin says, "The general implication for human health is an obvious one: An external agent can have an effect for a very long time. Given how long human generations last, the environmental exposures experienced by a pregnant mother can still have an effect 100 years later."

CAN PEOPLE SAFELY EAT CAT FOOD? 054

▶ IT MAKES MY FUR SHINY

Let's take a look at the ingredients in a typical can of cat food: meat by-products, chicken by-product meal, turkey by-product meal, ash, taurine. Nothing too horrible, but in general, these things don't constitute a healthy human diet, says Dawn Jackson Blatner, a registered dietitian with the American Dietetic Association. "That said, I'm fully confident that your body can handle kitty chow."

Your liver, kidneys, and skin do a terrific job of removing foreign substances from your body, especially mild ones like those found in cat food. "Technically, you could safely digest a baseball," Blatner says. But that doesn't mean you should. Perhaps the worst stuff in cat food is the high mineral content in the ash, but your body would clear that out quickly.

055 WHAT ABOUT THE FANCY EXPENSIVE STUFF?

▶ WHOLE (CAT) FOODS

Actually, the ingredients listed on the organic brands of cat food sound pretty tasty. Newman's Own canned beef formula uses only free-range beef from Uruguay, is 95 percent USDA-certified organic, and is chock-full of vitamins. Pass me a spoon, right? "Those are better," Blatner says, "but they, too, are developed with cat nutrition in mind and aren't formulated to keep humans healthy. It's OK to satisfy the occasional craving, but you shouldn't make it a staple of your regular diet. It's cat food for a reason."

056

I'M 26 AND MY VOICE STILL CRACKS. WHAT GIVES?

▶ SMOOTH TALKER

During puberty, boys' voices crack thanks to a surge in testosterone that causes the vocal cords to dramatically change shape and size, explains otolaryngologist H. Steven Sims of the Chicago Institute for Voice Care. As this is happening, the pitch of a boy's voice is in a frequent state of flux. "It can be hard to keep up with these changes and keep a smooth voice," Sims says. (This happens to girls as well, though not as much or as quickly, so girls are less prone to cracking.)

So why does a grown man's voice still crack from time to time? According to Sims, several factors could be involved. Weight gain can cause an increase in estrogen, which can raise the pitch of a man's vocal cords, making them more prone to squeaking.

Most often, though, extreme emotion or stress is the culprit. When the brain's emotional centers start firing with excitement, the electric impulses can sometimes spill over to the centers responsible for controlling speech coordination, causing the vocal cords to tighten or loosen unexpectedly and bring on the chirp. This means that even a woman's voice can crack, so at least everyone can share in the embarrassment.

DIY GUITAR HERO

Success in the music business these days requires more than a little technological know-how. *Popular Science* contributing troubadour Jonathan Coulton, who has skillfully used the Internet to build his music empire, gives aspiring rockers his formula for greatness.

"Chances are you've got a more advanced recording studio in your laptop than the Beatles had when they made *Sgt. Pepper's*, so record your music yourself," he suggests. "Then build an Internet home that can grow with your entourage. Skip the cookie-cutter MySpace stuff and get a full-fledged content-management system like WordPress or Drupal, which will allow you to expand: a blog, forums, photos, videos—all in one place that you control. And make sure it can support a digital music store so you can sell your own MP3s. Many prefer PayPal Micropayments, because the commission structure is better for small purchases than with iTunes or Amazon. If you don't know HTML or PHP, find a sucker . . . er, a fan to build it for you.

"Create merchandise on demand with CafePress, Spreadshirt, or Zazzle to avoid buying boxes of T-shirts that'll sit in your basement. There are still people who buy CDs, and for a low fee, CD Baby will store, sell, and ship your discs. It will also push your music to digital outlets like iTunes and Amazon MP3. When you're ready to play live, use Eventful.com, which lets people request a show in their town. Why slog from city to city in an old van unless you know you're going to sell some tickets?"

MOVE THAT MERCH

Coulton's also got some advice on do-it-yourself marketing. "Promote yourself on Twitter, broadcast live shows on Ustream, use Creative Commons licensing to encourage folks to make new content, such as music videos, with your music. Send out a million pieces of yourself to interact with potential fans. If they're out there, they'll find you—and hopefully sometime after that, give you money. Above all else, keep it simple and honest. Leave the 24-piece orchestra out of it (unless, of course, your band is a 24-piece orchestra)."

IS IT TRUE THAT I ONLY USE 10 PERCENT OF MY BRAIN?

▶ **YOUR FIRST-GRADE TEACHER WAS WRONG**

Historians have traced the earliest reference to this rumor back to the beginning of the twentieth century, when it was perpetuated by self-help gurus promising to expand people's mental abilities. But, like so many things hucksters have told us (sorry, that tonic isn't going to make hair grow on your chest), the brain claim is false. "There's no question," says Marcus Raichle, a neurologist and professor of radiology at Washington University in St. Louis. "You're using every little bit of this thing."

Even when you're sleeping or just sitting around watching TV, your brain is burning a surprising amount of energy for its size. Although your brain is about 2 percent of your body weight, it accounts for 20 percent of the total energy that your body consumes. Most of that energy, however, is used for tasks other than thinking.

Scientists know that most of your brain's energy is used for basic upkeep and communication among neurons. The rest, they speculate, might go toward preparing the brain to receive information by making predictions based on past experiences. For example, instead of scanning your entire fridge each time you go to grab some milk, you can reach directly for the shelf where you last left it—because your brain is working hard to remind you of its location and shoot your hand in that direction. This preprocessing helps us deal with the enormous amount of detail we encounter on a regular basis.

Anyway, you can be certain that all of your brain is working hard, even when you're not thinking hard. "We should back away from the notion that the only thing the brain is doing is sitting around waiting for something to happen," Raichle says. "Every piece of it is running full-tilt all the time."

10%

060
DO YOU USE MORE ENERGY WHEN YOU'RE THINKING REALLY HARD?

▶ **FEEL THE BURN(ING SYNAPSES)**

Need to lose some flab? Sit your big butt down with a math book—and get a workout. Kind of. The human brain is a 24-hour workhorse. While you're thinking, millions of neurons fire messages back and forth to each other and to the various tissues in the body. These neurons need fuel, consuming a full 75 percent of the blood sugar from the liver and 20 percent of the body's total used oxygen.

Here's how your neurons feed: Astrocytes—the cells near the capillary walls in your brain—suck energy-rich glucose from the bloodstream and convert it into a form that the neurons can soak up. The neurons then use it to fuel the production of neurotransmitters and, eventually, conscious thought. "The more energy an area of the brain wants, the more glucose that part of the brain will break down," explains neurologist Harry Chugani of the Children's Hospital of Michigan. "So yes, if you're thinking really hard and really struggling with your thoughts, the neurons in the frontal lobes of your brain will be burning a lot more glucose."

Simply put, to survive, your brain requires a tenth of a calorie per minute. Compare this with a walk to the doughnut shop, when your body burns approximately 4 calories a minute. Kickboxing zaps 10 calories a minute. And when you're hunched over a crossword puzzle? Your brain is blasting through a respectable 1.5 calories a minute. Pass the Cheetos.

061
WHO'S BETTER AT GIVING DIRECTIONS, MEN OR WOMEN?

▶ **THE ANSWER IS "YES"**

He says go straight for 15 minutes and turn east. She says drive past the school and turn right at the red house. Both sets of directions will get you to the same grocery store just as easily, but they embody the language barrier between the sexes that lurks behind many a front-seat argument.

Deborah Saucier, a professor of neuroscience at the University of Lethbridge in Canada, examined these phrasing differences in a 2003 study. She observed that, after examining a map and being asked how to get to various locations, women typically give directions that feature landmarks and left and right turns. Men, on the other hand, employ compass directions and distances measured in minutes or miles.

Some animals, such as homing pigeons, have extra iron in their heads that helps them turn toward the magnetic north pole. But men's internal maps, Saucier theorizes, most likely date back to our hunting ancestors. During a hunt, men would stray far from home and into unfamiliar territory to bring down wild animals. They may have relied on tracking the position of the sun and their innate orientation skills to find the most direct route home.

Meanwhile, prehistoric women, who gathered more sedentary food, probably regularly found their way to and from the most bountiful and nutritious plants using landmarks. In a study at the University of California at Santa Barbara, evolutionary psychologist Joshua New tested this theory in a farmers' market. After a single tour of the market, women could more accurately point to food stalls they had visited, noting, in particular, the locations of foods with high energy content. With food stalls as landmarks, women knew their way around better than men.

So neither way is better; they're just different. To compensate for the gender differences, Saucier suggests giving disoriented people both male- and female-oriented instructions. "People get a lot less lost that way," she says.

ARE MEN OR WOMEN MORE LIKELY TO BE HIT BY LIGHTNING? 062

PUT DOWN THE GOLF CLUBS

The numbers tell the story: Of the 648 people killed by lightning in the U.S. from 1995 to 2008, 82 percent were male. And as much as we were hoping to uncover a concrete biological cause—extra iron in the male cranium, perhaps, or the highly conductive properties of testosterone—it turns out men are . . . just kind of stupid. "Men take more risks in lightning storms," says John Jensenius, a lightning safety expert with the National Weather Service.

Men are less willing to give up what they're doing just because of a little inclement weather, Jensenius says, and will continue to engage in pastimes that make them vulnerable, such as fishing, camping, and golfing. Recreational or sports-related activities are involved in almost half of all deaths caused by lightning.

Peter Todd, a behavioral psychologist at Indiana University, suspects the difference between the sexes boils down to the basic risk-versus-reward systems that have been part of our biological wiring for thousands of years. For women, Todd explains, the ancient biological priorities are to protect one's reproductive role and to care for offspring, which outweighs any inclination to attract potential mates by exhibiting bold behavior.

But for men, Todd says, the risk of getting struck by lightning could be outweighed by the reward of proving to other men—and potential female mates—that they're not afraid of getting struck by lightning. This is particularly true for young men, who have the most to gain by impressing others, thereby raising their status as attractive, daring, healthy mates in the dating pool. And then, zap!

CAN MEN PRODUCE BREAST MILK? 063

MR. MOM

Cases are rare, but yes. The impetus for milk production begins in the pituitary gland, a pea-sized organ on the underside of the brain. Normally, the pituitary churns out signals that keep our bodies functioning and producing the correct types and levels of hormones. However: "This gland is susceptible to disease, just like any organ in the body," says Charles F. Abboud, who specializes in endocrinology at the Mayo Clinic in Rochester, Minnesota. And oh, the horrors that befall the man whose pituitary gland starts misbehaving.

One of the side effects of a wonky pituitary can be the production of a pregnancy hormone, prolactin, a compound that encourages the breast tissue to produce milk. For most men, extra prolactin would shut off the reproductive system, limiting their ability to produce sperm. In a select few, however, it can also lead to, as Abboud delicately puts it, "unusual breast secretions."

064 HOW MUCH CAN A HUMAN BODY SWEAT BEFORE IT RUNS OUT?

▶ **RUNNING ON EMPTY**

It all depends on the size, physical fitness, and hydration of the person in question, but it's possible to sweat an awful lot before heatstroke sets in and we pass out. After all, there are about three million sweat glands on the human body (the highest concentration is on our palms), and the average person who's working out aggressively perspires around 1.5 to 3.2 pints (0.7–1.5 liters) per hour. Theoretically, if you were attached to a treadmill and pumped full of liquids, it's possible to keep sweating forever.

Very active people sweat about 3.2 to 3.8 pints (1.5–1.8 liters) an hour, while a triathlete can produce nearly 1 gallon (4 liters) of sweat in the same time.

During the Ironman Hawaii, competitors perspire some 4 gallons (about 15 liters) as they run a marathon, swim 2.4 miles (3.9 km), and bike 112 miles (180.2 km). Lawrence Spriet, an exercise physiologist at the University of Guelph in Ontario, says that after one loses 3 to 5 percent of one's body weight, the sweating process begins to slow down.

Lawrence Armstrong, an environmental and exercise physiologist, has proved that the human body continues sweating no matter how dehydrated it is. As long as the hypothalamus sends nerve impulses to the sweat glands, we'll perspire. If sweating stops, then something is terribly wrong.

065 WILL I DIE IF I RUN OUT OF SWEAT?

▶ **KEEP A COOL HEAD**

The whole point of sweat is to keep the body cool. This is pretty vital: If your core temperature goes above 104°F (40°C), the body begins to overheat to the point where its proteins denature. When this happens, "membranes of the tissues lose their integrity, and things leak out," Lawrence Spriet says. The intestines can discharge bacteria into the bloodstream, and the body goes into shock. By then you'd probably be unconscious, possibly even in a coma. But while people do in fact die of overheating, it's very unlikely to be due to a sweat shortage. As Lawrence Armstrong points out, even in extreme cases it's impossible to sweat out all the water in our bodies: "People don't shrivel up until they are dead."

PREHISTORIC PODIATRY

Hard to know, says Will Harcourt-Smith, an expert on early human fossils at the American Museum of Natural History in New York. "Some infections leave their mark on bones. Athlete's foot is not one of those infections. But if we make some logical assumptions, we might be able to make a good guess."

Athlete's foot is a fungal infection of the skin—typically by fungi of the *Trichophyton* genus—that causes skin to scale, flake, and itch. Which makes us ask: Did cavemen even encounter this fungus? "The fungus that causes athlete's foot was definitely around back then, and probably much earlier," says Tim James, who specializes in fungi evolution at the University of Michigan. "Like all fungi, it thrives in moist, unhygienic environments, which is why most people pick it up in locker rooms. I don't imagine that a caveman's dwelling, with a dirt floor covered in animal remains, was a very sterile place."

But just walking around in fungus doesn't cause athlete's foot. Cavemen would have had to have worn shoes. "It turns out that athlete's foot is a disease of shod populations," says Bob Neinast, the lead blogger for the Society for Barefoot Living. "Anyone can pick up the fungus, but the thing to keep in mind is that it grows really well in a warm, dark, moist environment. That's the inside of a shoe." People who go barefoot,

Neinast says, rarely get athlete's foot, most likely because exposure to fresh air keeps their feet too dry for the fungus to take hold and multiply.

Which leads us to ask: Did cavemen go barefoot? "Within around 10,000 years ago, people had lovely shoes," Harcourt-Smith says. Our ancestors might have moved out of caves and into villages by that time, he notes, but their footwear was usually pretty basic, consisting of leather wrappings sometimes stuffed with grass for insulation during cold weather. "If the shoes got damp and the person wore them often enough, that could have encouraged athlete's foot," he says.

Even the worst case of athlete's foot wouldn't have killed a caveman, but it could have impaired his quality of life. "If the irritation gets bad enough, it will stop you in your tracks," says Cody Lundin, an outdoor survival–skills instructor who has gone barefoot for 20 years. "That would be unacceptable for a hunter population." Without antifungal sprays or creams, how would they have fought the burn? They might have been able to cook up a remedy. "If you take the green parts of a juniper plant and boil them, the mix makes a wonderful fungicide that will work on athlete's foot. Indigenous people might have used it," Lundin says. "Works great on jock itch, too."

067 WHY DO BOYS AND GIRLS FIGHT DIFFERENTLY?

▶ BATTLE OF THE SEXES

Science is still some way from explaining why boys throw punches and girls pull hair on the playground. In the boxing ring, though, researchers are making progress—at least when the combatants are fruit flies.

Edward Kravitz, a professor of neurobiology at Harvard Medical School and the Don King of fly fights, has identified a gene that controls the fighting tactics of male and female fruit flies. To instigate the same-sex battles, Kravitz offers lavish prizes: yeast for the females and, for the males, the privilege of courting a headless female. "We won't get into social commentary with that fact," he jokes.

As you might expect, the male fruit flies fight fiercely. They lunge and then rear up to drop blows with their forelegs. "If you watch in slow motion, they just flatten their opponents," Kravitz says. Females choose to shove and head-butt, a daintier but equally effective approach.

But are the different fighting styles learned behavior, or are they hardwired in the flies' DNA? To find out, Kravitz transplanted the male version of a gene previously associated with gender-specific behaviors, such as male courtship, into females and put the female version in males. After the gene swap, the males fought like females and the females used male techniques, providing the first evidence of gene-controlled gender-specific aggression in fruit flies.

Kravitz's research doesn't translate smoothly to humans, however, because we don't yet know of a gene corresponding to the one he swapped in flies. Also, scientists generally agree that different levels of testosterone exposure early in life are probably responsible for the aggression-related gender variation in humans. Still, Kravitz thinks his findings will help neurobiologists understand how complex behaviors like aggression get wired into the nervous system—without anyone getting pummeled in the ring.

DO MUSIC LESSONS MAKE KIDS SMARTER? 068

LISTEN TO YOUR TIGER MOM

Several studies published in the past few years involving children ages 4 to 15 have strengthened the theory that music lessons have a positive effect on kids' brains. The first, by the Chinese University of Hong Kong, looked at 90 boys between the ages of 6 and 15; half were in the school's string orchestra, and the rest had no musical training. Another study, by the University of Toronto, enlisted 144 6-year-olds and randomly assigned them to a year of piano lessons, voice coaching, or nothing.

The researchers discovered that lessons on a musical instrument can boost mathematic ability and overall IQ. Not surprisingly, the longer a student sticks with it, the greater those improvements.

069 MY KID DIDN'T LAST LONG IN MARCHING BAND. DID HE STILL GET SMARTER?

EVERY LITTLE BIT HELPS

Scientists at McMaster University and the Rotman Research Institute in Toronto revealed that as little as four months of music lessons cause noticeable improvements during brain development.

The researchers' study followed a handful of aspiring 4- to 6-year-old musicians over the course of a year, measuring the patterns of neuronal activity in each participant's brain.

When the scientists compared these young Mozarts with a control group, they found that music students' brains were developing differently. For example, it appeared that music instruction improved an information-processing area of the brain associated with attention. Unlike their peers, the nascent musicians' general memory capacity (measured by how easily subjects could memorize strings of numbers) increased over the course of the year.

Before you start kicking yourself for having given up on the tuba too early, take comfort: It appears that even quitters reap some benefits. Studies have shown that instead of losing their abilities wholesale, children who stopped their lessons retain some of the skills engendered by their musical training.

070 I'VE HEARD THAT THE Y CHROMOSOME IS DOOMED. ARE MEN ON THE WAY OUT?

▶ FIRST, SOME BACKGROUND

Humans store their genes in 23 pairs of chromosomes, 22 of which are identically matched. The 23rd is a two-sided biological coin—twin Xs mean you're female; an X and a Y, male. Chromosome pairs often trade bits of DNA in a process called recombination, the purpose of which is to keep genes functioning properly.

Talk of men's path toward extinction began in the late 1990s, when it was discovered that the human Y chromosome, which is stumpy compared with the X, does not share enough genetic material with the X to practice recombination. Left without a way to renew damaged genes, the Y would continue to degrade and would eventually disappear, geneticists announced. They slapped an expiration date on the male half of the species of sometime in the next 5 to 10 million years.

▶ CHIMP PREDICAMENT

To get a perspective on this prediction, scientists looked to our closest genetic relatives—the chimps. Because humans and chimpanzees shared a common ancestor six million years ago, geneticist David Page of the Whitehead Institute for Biomedical Research in Cambridge, Massachusetts, studied how the chimp Y chromosome and its human Y counterpart have evolved differently in the intervening years. What he found surprised him: The chimp Y chromosome is far more degraded than the human Y chromosome.

Page and his colleagues speculate that chimps' promiscuity—females mate with multiple partners—has led to enhancement of the Y genes that produce sperm, to the detriment of other genes. Among chimps, "there are sperm wars going on. Each male is trying to pass his own genes down," says Jennifer Hughes, who coauthored the study, published in *Nature*. Neglected, the chimp Y chromosome's nonreproductive genes have declined.

▶ SO WHAT DOES THIS MEAN FOR US?

The degradation of the chimp Y reveals something relevant to humans. The Whitehead Institute scientists think that although the human Y chromosome also lost genes at first, in recent eons it has been relatively stable. The human Y has eluded the chimp Y's fate, they suggest, because humans are largely monogamous. Human sperm don't face the same competition as chimps', so there isn't as much pressure on the human Y to produce good sperm.

Not all geneticists are convinced that the human Y has stopped deteriorating. Jenny A. Marshall Graves of the Australian National University in Canberra, for one, still believes that the chromosome's days are numbered. "The human Y has been degenerating since it was born, 300 million years ago," she says. And so the controversy continues. Rest assured, though; the Y—and the guys—will be around for a while.

071
WHY DO I FEEL SLEEPY AFTER I EAT?

▶ **FOOD COMA**

The glucose from food causes your body to reduce production of orexins, a class of proteins that is believed to keep us alert. Orexin-producing neurons, which are found in the hypothalamus region of the brain, have been known since 2003 to be affected by glucose levels, but it wasn't known to what extent those levels—and the subsequent suppression of orexins—interfere with alertness. Recently, Denis Burdakov and his fellow researchers at the University of Manchester in England reported that even a small increase in blood-glucose levels significantly lowers the orexin-mediated neuron activity in the brain, causing sleepiness. The strong dependence of alertness on glucose levels was surprising, although, Burdakov says, the discovery makes perfect evolutionary sense. "It would be advantageous for animals to stop energy-expending behaviors after they obtained their food, in order to make the calories last as long as possible," he explains.

PASS ON THE PASTA

But all foods are not created equal. According to Burdakov, meals rich in carbohydrates or fats elevate blood-glucose levels and make you more tired; meals higher in protein are less likely to have that effect. So if you want to make it through the workday without nodding off at your desk, eat some meat for lunch.

WHY IS YAWNING CONTAGIOUS? 072

YOU'RE GETTING SLEEPY

Go ahead, admit it—you're yawning right now. That's OK, we forgive you. We know it's not because you're bored. It's just that seeing other people yawn, reading about yawning, or even just thinking about it can make you yawn. It may be annoying, but it could be a sign that your brain is in tune with those around you.

Picking up another person's yawn may be an empathetic reflex, says evolutionary psychologist Gordon Gallup of the State University of New York at Albany. When you see someone yawn, neurons in your brain fire and cause you to "feel" what that person is experiencing, commanding you to perform the action even if you don't actually feel the need.

073 SO WHY DO WE NEED TO YAWN AT ALL?

EARLY TO BED

Scientists have yet to figure this out. Some say it signals boredom, whereas others have suggested that it balances carbon dioxide and oxygen levels in the blood. Most recently, Gordon Gallup theorized in a study that yawns cool our brains so that they run efficiently, similar to running a fan in a computer. He found that, contrary to common perception, cooling the brain with a yawn actually keeps people from nodding off, a handy trait when being stalked by predators.

Alternatively, E.O. Smith, an emeritus professor of anthropology at Emory University, contends that yawning may have encouraged our ancestors to go to bed early. "Contagious yawning didn't arise in a vacuum on a Saturday afternoon," Smith says. Spreading a case of the yawns around a bonfire, he posits, may have prompted night owls to climb up into their comfy tree beds with their sleepier brethren, out of danger from stealthy predators.

()74
WHAT IS THE EVOLUTIONARY PURPOSE OF TICKLING?

BONDING MOMENT

You probably know that you can't tickle yourself. And although you might be able to tickle a total stranger, your brain also strongly discourages you from doing something so socially awkward. These facts offer insight into tickling's evolutionary purpose, says Robert R. Provine, a neuroscientist at the University of Maryland and the author of the book *Laughter: A Scientific Investigation*. Tickling, he says, is partly a mechanism for social bonding between close companions and helps forge relationships between family members and friends.

Laughter in response to tickling kicks in during the first few months of life. "It's one of the first forms of communication between babies and their caregivers," Provine says. Parents learn to tickle a baby only as long as she laughs in response. When the baby starts fussing instead, they stop. The face-to-face activity also opens the door for other interactions.

Children enthusiastically tickle one another, which some scientists say not only inspires peer bonding but also might help hone reflexes and self-defense skills. In 1984, psychiatrist Donald Black of the University of Iowa noted that many ticklish parts of the body, such as the neck and the ribs, are also the most vulnerable in combat. He inferred that children learn to protect those parts during tickle fights, a relatively safe activity.

CHILD'S PLAY

Tickling while horsing around also may have given rise to laughter itself. "The 'ha ha' of human laughter almost certainly evolved from the 'pant pant' of rough-and-tumble human play," says Provine, who bases that conclusion on observations of panting among apes that tickle each other, such as chimpanzees and orangutans.

In adulthood our response to tickling trails off around the age of 40. At that point, the fun stops; for reasons unknown, tickling seems to be mainly for the young.

WHY DO FINGERNAILS GROW FASTER THAN TOENAILS?

▶ TIME FOR A MANICURE

If you find that you clip your fingernails far more frequently than your toenails, it's not necessarily a reflection of poor personal hygiene. Fingernails grow about one-tenth of an inch (2.5 mm) per month, about twice as fast as toenails.

Dermatologists and scientists have yet to determine the reason for this disparity in growth rate, but one theory suggests that your hands benefit from better blood circulation—and thus, a more robust supply of oxygen and growth-fueling nutrients—because they are physically closer to your heart than your feet. It's also possible that persistent minor traumas, such as typing or scratching, actually stimulate nail growth, whereas our toes enjoy a virtually trauma-free existence inside socks and shoes.

Some evidence supports the trauma theory: Fingernails on the dominant, most frequently used hand grow the fastest. And the shorter a finger, the slower its nail grows, perhaps because its longer neighbors shield it from the constant banging around.

If you bite your nails to keep their length in check, you could be working against yourself. "Minor occasional nail-biting might make the nail grow faster," says Paul Kechijian, a dermatologist and former chief of the nail division at New York University Medical Center. If you're flexible enough, a nibble every once in a while might help your toenails catch up.

076
WHY DO CELL PHONES TURN PEOPLE INTO SUCH CRUMMY DRIVERS?

▶ **HANG UP AND DRIVE**

Driving safely (and not being a jerk) requires full awareness of your surroundings. This demands the brain's full attention, which is often sidelined as we overwhelm our frontal lobes' processing power with the ultra-distracting cell phone, trying to do too many tasks at once.

"Multitasking" is a term that originally referred to a computer's ability to execute several commands concurrently. Naturally, we assume our brains can do the same thing. But many researchers agree that our alleged modern ability for gaming, downloading, texting, and talking simultaneously is just a myth. Technically, our big human brains can't even do two things at the same time without paying a price.

The multitasking myth, says cognitive researcher Jordan Grafman of the National Institute of Neurological Disorders and Stroke, stems from the ability of part of the frontal lobe to toggle back and forth among simple tasks in as little as a few hundred milliseconds and among complex tasks in only a few seconds. It may seem like doing two things at once, but even these tiny lapses create delayed response times that can lead to car accidents. What's more, Grafman says, the frontal lobe is also responsible for your ability to observe and reflect on your surroundings. Cell phones degrade this ability—because you can't see the person you're talking to, the phone creates an extra demand on your brain (which probably explains why people on the street bray into their phones, despite your giving them the evil eye). Finally, according to a recent University of Utah study, phoning while driving doubles the likelihood of rear-ending the car in front of you, even if you're using a hands-free headset—the lack of visual cues, compared with just speaking with a passenger, is too taxing.

077
WHY DO
DUMB GUYS
SEEM TO IMPRESS THE LADIES?

▶ **BRAINS VS. BALLS**

Although many jilted brainiacs might beg to differ, there's no concrete evidence that women are more attracted to dumb men. Yet the same might not be true for some of our mammalian cousins. Consider, for example, the bat. After gathering available brain- and testis-size data for 334 species of bats, evolutionary biologist Scott Pitnick of Syracuse University found that males with the biggest cranial capacity were likely to have the smallest testicles, and vice versa.

Why does this trade-off occur? Both brain and testis tissue are physiologically demanding to grow and maintain. Because bats' high metabolism and near-constant movement don't leave them much energy to spare, many species have evolved to favor one organ over the other. Testes tend to be the largest in species wherein females are especially promiscuous—size (and, thus, sperm count) has the advantage when sperm from several males are competing in the reproductive tract.

"The question is, do you get your genes to the next generation by being clever or by getting busy?" Pitnick says. Because humans live slower, easier lives than bats, he adds, they haven't faced the same evolutionary dilemma. This could mean that, for humans at least, having a big brain is a more important evolutionary advantage than producing a lot of sperm. "People ask me if this means Albert Einstein had tiny testes," Pitnick says, "and that's not necessarily true."

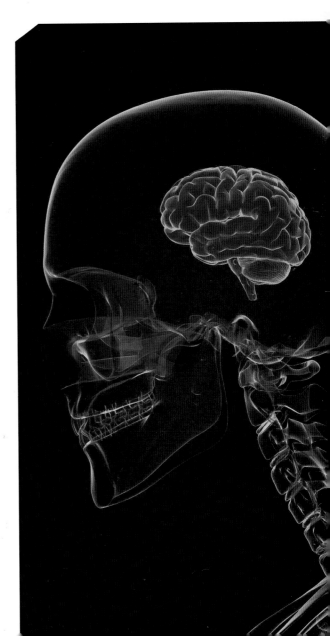

078 WHAT'S THE **WORST** SOUND IN THE WORLD?

▶ MAKE IT STOP

The grinding whine of a dentist's drill? A beginning violin student? Not even close, say researchers who set out to determine the world's most repulsive sound.

The group, led by acoustic engineer Trevor Cox of the University of Salford in England, uploaded a selection of 34 sounds to their website, Sound101.org, and asked users to rate each sound on a scale of one (not bad) to six (plug-your-ears-and-cry horrible). More than 1.6 million votes later, the answer was clear across all age groups and cultures: There's nothing worse than the sound of someone else vomiting.

To create what Cox calls the proper "slopping sound," the team had an actor make yacking noises while he dumped baked beans into a bucket. That vile audio snippet beat out strong contenders, such as squealing microphone feedback and an electronic version of fingernails clawing a blackboard. "We're looking at what psychologists call a disgust reaction, which is a survival mechanism," Cox says. "If someone's coughing or vomiting, they're carrying disease, and thus we avoid it."

079 IS THE DISGUST REACTION **INSTINCTUAL OR CULTURAL?**

▶ THAT'S SICKENING

Although Trevor Cox believes that culture plays a role in what sounds we find repulsive, the tendency to be repulsed by the sounds of disease is probably an instinct, which may have evolved as a mechanism to avoid catching an illness from a neighbor. In 2004, Val Curtis at the London School of Hygiene & Tropical Medicine ran a study using disgusting photos rather than sounds and also concluded that disease is universally the most revolting characteristic. For the record, the blue ribbon went to a photo of a person's gums infested with parasite eggs—which might even be gross enough to elicit genuine vomiting.

SILENT BUT DEADLY

If you've been staying up late pondering this question, you're in luck. For their senior project, two Cornell University computer-engineering whizzes recently built a machine that does just that. After learning in class how Breathalyzers work, Robert Clain and Miguel Salas assembled a fart detector from a sensitive hydrogen sulfide monitor, a thermometer, and a microphone and wrote the software that would rate the emission. A "slight perturbation in the air" near the detector sets it to work measuring the three pillars of fart quality: stench, temperature, and sound. Temperature, Clain explains, is critical. The hotter a fart, the faster it spreads. "It beeps faster if it's a high ranker, and a voice rates it on a scale of zero to nine," he says. "If it ranks a nine, a fan comes on to blow it away. It even records the noise so you can play it back later." After a few months of construction, they began field tests. "Well, the sample data wasn't collected from the entire school, but we definitely tested it," Salas says.

081 HOW CAN FART-MEASURING TECHNOLOGY MAKE THE WORLD A BETTER PLACE?

STENCH-O-MATIC

In fact, this contraption could actually have uses outside of fraternity houses, Robert Clain says, as a biosensor for harmful hydrogen sulfide–producing bacteria in hospitals. Or dentists could use it to measure oral malodor. They've also received some interest from doctors with four-legged patients. "You can test the health of livestock through the quality of their farts," Miguel Salas says. "Smell and sound can tell you a lot about their bowel movements."

When it came time to present the invention in class, though, Clain and Salas had to test their detector by making raspberry sounds and breathing on it—human exhalations contain enough hydrogen sulfide to trigger the sensor. "It's hard to fart something really smelly on command," Clain laments. "Besides, it provided a nicer atmosphere for those around us." Still, their professor saw fit to award the project a well-deserved A.

082

IF I ATE LAB-GROWN HUMAN TISSUE, WOULD I BE CONSIDERED A CANNIBAL?

▶ AN ALTERNATIVE TO SOYLENT GREEN

If you're looking to indulge in the other, other white meat but can't stand the idea of society branding you a cannibal for your experimentation, this might be the loophole you're looking for. And there are plenty of dishes to choose from at a university near you.

Since winning a *Popular Science* Best of What's New Award in 2006 for the world's first artificially grown human-tissue replacement, Anthony Atala, the director of Wake Forest University's Institute for Regenerative Medicine, has been busy trying to cultivate more than 20 types of human tissue, including liver, kidneys, lungs, heart valves, skeletal muscle, erectile tissue, and bone. These tissues, grown from cells harvested from patients, are as good as what you were born with. "In tissues that have been successfully implanted in humans, testing shows no discernible differences between the native tissue and the lab-engineered tissue," says Atala (who declined to speculate on what those tissues might taste like). In other words, a kidney grown in a dish is biologically identical to one freshly plucked from a living person.

Whether eating that lab-grown kidney would constitute an act of cannibalism is much less clear-cut, and that's primarily because the common definition of "cannibal" isn't much more specific than "a human being who eats other humans."

"What grosses us out about cannibalism is that it involves using other people as a means of nutrition or biologically integrating some other human being into oneself," says Jonathan Moreno, a professor of medical ethics at the University of Pennsylvania's Center for Bioethics. "There's also a certain nauseating sensation that detached body parts elicit, because you know that they once belonged to another human." But because artificial organs are essentially a few cells multiplied many times over and shaped by a protein scaffold, the connection to another person is much weaker, Moreno explains. "So I guess I'm not persuaded that eating them would count as cannibalism."

That said, Moreno considers the whole thing a bad idea. "Compared to meat derived from lower animals, I don't see the nutritional benefit for humans," he says.

083 SO I WON'T BE EATING AT THE LONG PORK CAFÉ ANY TIME SOON?

▶ PEOPLE, PEOPLE WHO EAT PEOPLE

It's extremely unlikely that anyone is going to start marketing human steaks in the near future, or ever. Even leaving the ethical questions aside, serving lab-grown humans is just bad business. "Let's say a company grew more than a slab of meat—a whole limb," Jonathan Moreno says. "They couldn't advertise it as a human body part, only that it was related to

human tissue, and that might not be exciting enough to make someone think that they're breaking a taboo. This wouldn't be cannibalism, but it objectifies body parts and could be seen as commodifying a good that unacceptably resembles a human body part. And that could get you into some legal issues." So, adventurous eaters, sorry, but we suggest keeping lab-grown tissue in the operating room and off the dinner table.

084
WHY DOES SPITTING IN MY SCUBA MASK KEEP IT FROM FOGGING UP?

▶ SPIT-SHINE

Fog forms on the inside of your goggles because of the temperature difference between the cool outside water and your 98.6°F (37°C) face. Just as water droplets condense on the outside of a glass of ice water, your chilly lenses cause water in the air trapped inside your goggles to condense into tiny, mistlike beads that gather on the inside surface of your lens and scatter light every which way, making it impossible to spot all those colorful reef fish.

Unless you're swimming in water that's body temperature (or warmer), your best bet is to prevent the condensation from forming beads. Most anti-fog goggle treatments coat the lens with hydrophilic (water-loving) chemicals that pull the water flat to the lens's surface. Any condensation takes the form of a clear, thin, continuous film across the surface of the lens that doesn't diffract light as much as the mosaic of spherical beads, says Chris Ryser, president of Opto Chemicals, the maker of Zero Fog sprays and creams.

The mucus in your saliva works much the same way, says Dennis Lopatin of the University of Michigan. Scientists don't know for certain, but it probably also lowers the surface tension of the lens, making it even more difficult for water to form beads. Unfortunately, the spit treatment doesn't last as long as the synthetic stuff, so you may need to reapply frequently.

085

HOW FAST WOULD SOMEONE HAVE TO SPRINT TO RUN ON WATER?

▶ **MIRACULOUS MARATHON**

When Jamaican sprinter Usain Bolt set the world record in the 100-meter dash at the 2008 Beijing Olympics, he was zipping along at an astonishing 23 miles per hour (37 kph). But the world's fastest man would have to move at highway speeds to take his show from turf to surf.

According to John Bush, a professor of applied mathematics at the Massachusetts Institute of Technology, an adult human would have to sprint at nearly 70 miles per hour (112.6 kph) to avoid falling through the water. But it's not just about moving fast. When basilisk lizards run over water, they're actually running on top of air pockets created as their feet slap the surface. Doing this, however, requires significant leg strength to pull their feet up before those pockets can dissipate. A human would need leg muscles 15 times stronger than normal to accomplish this feat.

If you're planning a more leisurely jaunt, try something like the "water-walking shoes" designed by Wavewalk in Massachusetts. The kayak-like boots, which measure 6 feet (1.8 m), will allow you to float across the water like a snowshoer. Just don't expect to set any sprinting records.

086
IS IT POSSIBLE TO DO MORE THAN TWO BACKFLIPS ON A MOTOCROSS BIKE?

▶ HANG ON TIGHT

James Riordon, an avid skateboarder, studied physics at the University of Maryland and works for the American Physical Society. He used his varied talents to explore this vital question.

"My son and I are huge fans of freestyle motocross superstar Travis Pastrana, and after he pulled off the first-ever double backflip in the 2006 X Games, we wondered how many more flips he might be able to do. These guys push the limits, but we figured there must be a physical limit to how many backflips one can do on a motocross bike. So I applied some Newtonian physics to the trick to figure it out.

"The average rider and bike together weigh about 330 pounds (150 kg), and the highest they can jump (without flipping) is about 36 feet (11 m). But it takes energy to rotate the bike, which a rider pays for in lost height.

"I did some calculus to figure out the best way to budget the rider's energy between jumping and rotating, and it turns out that a 50-50 split is best. This means that the rider will launch 18 feet (5.5 m) high and will be airborne for 2.12 seconds. In this amount of time, spinning at 1.67 rotations per second, there's no way for anyone to do more than four backflips.

"Remember, though, that this is the absolute limit on these bikes, and full rotations are a must. If the rider rotates only, say, three and a quarter times, he's going to crush his front fork and end up in the hospital. That said: Travis, you could easily do three. Just attempt it into a foam landing pit first."

CAN A SNOWBOARDER BEAT A SKIER IN A DOWNHILL RACE?

087

COLD WAR

From a physics standpoint, the factors affecting speed (friction and drag) are about the same for snowboards and skis. The main thing that prevents snowboarders from keeping up with skiers is that boards are more likely to buck the rider at high speeds.

Snowboards are excellent for carving hard turns, but a board going straight and flat-bottomed—which is the fastest way to travel down the mountain for both skis and boards—will twitch to one side or the other. Snow is bumpy, so you need to make gentle adjustments to stay in a straight line at high speeds, but you must lean your center of mass way out over the side of the board to get up on an edge to execute a corrective turn.

That's a big adjustment, and it increases the chance that you'll catch too much edge and slow down (or fall on your butt) before you reach your top speed. If not for this, snowboarders could go just as fast as skiers.

Skiers, on the other hand, can widen their stance for more stability when riding flat-bottomed. (Contrary to popular belief, long skis and snowboards don't slide faster than short ones: They're just more stable and allow you to reach faster speeds without wiping out.)

088

DO SNOW CONDITIONS CHANGE THE ODDS?

POWDER PLAY

It's a good bet that a snowboarder could probably blast by a skier in fresh powder that's at least 6 inches (15 cm) deep. The fluffy stuff essentially acts like a liquid in which the board behaves more like a surfboard—you have to generate some lift so that you don't sink into the snow. A broad snowboard can rise up on top of the snow with relative ease and basically skim down the mountain. Skinny skis, on the other hand, need to take a much steeper angle of attack to rise out of the snow. This creates more drag and exerts more force backward, which slows down the skier. Definitely bet on a snowboarder in those conditions.

089 WOULD TWO PAIRS OF IDENTICAL TWINS HAVE IDENTICAL KIDS?

▶ PEAS IN A POD

Sets of identical twins actually pair up more often than you might guess, forming what are known as "quaternary" marriages. Fortunately for the rest of us, the result is not an army of clones.

In fact, the chance of the kids of two pairs of twins being identical is zero, says geneticist Rob Martienssen of Cold Spring Harbor Laboratory in New York. The only way to produce identical people is to have a single fertilized egg split into two identical embryos in the womb. But, because the children of a quaternary marriage are born of separate wombs, they can't undergo this twinning process.

Instead, as with any child with nonidentical parents, each embryo receives a random assortment of genes from mom and dad. And even though in a quaternary marriage the pairs of moms and dads are genetically identical, there is no realistic chance that these gene selections—for characteristics such as eye color and height—would be the same for each child. Thus, the offspring would be no more likely to look alike than a pair of siblings produced by a non-twin couple.

All this assumes, of course, that the twins are still truly identical in the first place. As people age, small chemical groups are added to 1 percent of their genome, a process called methylation. These groups don't rewrite your genetic code, but they can affect how genes make proteins.

Although scientists have yet to observe the inheritance of methylation patterns in humans, studies have revealed that this can happen in mice. So by the time twins have their own children, their DNA blueprint is probably no longer identical anyway. In other words, even identical twins aren't really identical.

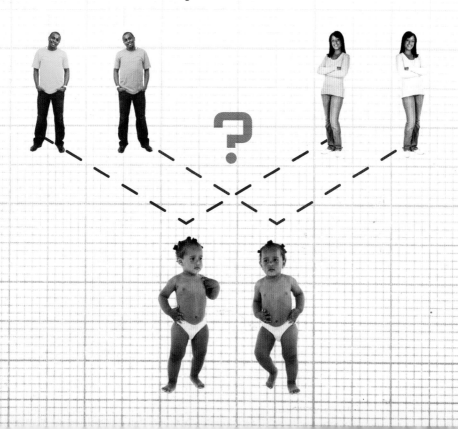

DO BEAUTIFUL PEOPLE ALWAYS HAVE BEAUTIFUL BABIES?

▶ STAR POWER

The children of movie stars might inherit their parents' wealth and enviable connections, but they're not guaranteed to get the sizzling good looks. One study suggests that the reproductively fittest members of a species—which roughly translates to the sexiest—might actually produce the least fit offspring.

SON, YOU TAKE AFTER YOUR MOTHER

It seems the genes that drive mating behavior, called "sexually antagonistic" genes, are to blame. Though beneficial to the mating success of one sex, one of these genes can squash the reproductive chances of the other, explains Alison Pischedda, who was a graduate student at Queen's University in Ontario when she led the study.

If highly fit males pass on masculine genes to daughters, or fit females pass on feminine genes to sons, Pischedda found, the offspring will be less successful in finding mates. Surprisingly, the children of parents that were considered the most beautiful of each sex—the study looked at fruit flies, but yes, some flies are more attractive than others—were the least fit of all the flies.

Although scientists don't know exactly how these genes work in humans, the study offers an explanation for how a beautiful couple might produce unattractive children. "Consider facial features," Pischedda says. "A masculine, square jaw, which is handsome in a man, is not very beautiful in a daughter."

091 WHY IS IT SO HARD TO WAKE UP IN THE MORNING?

NOT A MORNING PERSON

It's not necessarily laziness that makes people hit the "snooze" button in the morning. Most likely, your body clock is mismatched with the demands of your life.

Your clock is controlled by the suprachiasmatic nucleus, a part of the brain that controls the body's biological rhythms. But, says Jean Matheson, a sleep-disorders specialist at Beth Israel Deaconess Medical Center, these preset natural rhythms often don't align with daily realities—work or school start times cannot be adjusted to fit a person's sleep schedule. People who have trouble crawling out of bed probably have an inner clock set to late wake-up and sleep times, a condition known as phase delay.

BACK ON TRACK

It is actually possible to adjust your phase-delayed body clock, Matheson says, but at a price: no more sleeping in on weekend mornings. "When people sleep late on weekends, they revert to their natural phase-delayed rhythm," she explains. This makes it harder to wake up early on weekdays.

You can train yourself to wake up earlier, Matheson says, by setting your alarm 15 minutes earlier each day (and heeding its call).

Exposure to artificial light in the evening can also cause phase delay. The brain is very sensitive to light, and too much of it just before bed—from computer screens, televisions, or bright reading lights—can trick the brain into thinking it's daytime.

If you find it difficult to adjust your sleep habits, there's some good news. Scientists at the University of California at Irvine recently discovered that a single amino acid regulates your internal clock. One day, says pharmacology professor Paolo Sassone-Corsi, this research could translate into a drug that controls the brain's sleep cycle.

092 WHAT DO MY EYES DO WHILE I'M SLEEPING?

EYES WIDE SHUT

When your head hits the pillow and you lower your lids, your eyes still function. "But they can only sense light versus dark," says physician Michael Breus, a clinical psychologist and a fellow of the American Academy of Sleep Medicine. This explains why a bright light or the sunrise often wakes a person up.

Still, the eyes don't send image information to the brain during sleep. In fact, the connection between your eyes and your brain needs to reboot when you wake up. "It takes 30 seconds for the brain to recognize that you are awake," Breus says. "That's why you don't see complete images right away."

Your eyes do zip around during rapid-eye movement (REM) sleep, but they aren't sending any visual data to the brain, even though studies have shown that the visual cortex—the part of the brain that processes images—is active. Scientists believe that this activity marks a memory-forming or memory-reinforcing process, perhaps locking in your recollection of what happened during the day.

093
WHY DO SOME PEOPLE DREAM MORE THAN OTHERS?

▶ **AND THEN THE ALIENS TOOK ME AND ABRAHAM LINCOLN ON THEIR SPACESHIP . . .**

Unless you're an unlucky teen in *A Nightmare on Elm Street*, you probably dream the same amount as everyone else: upward of two hours a night, according to the National Institutes of Health. The brain's sleep cycle, however, makes it extremely difficult to remember dreams. "During sleep, our memory systems are completely shut down, and we're basically living on a self-erasing tape," explains neurologist Mark Mahowald, director of the Minnesota Regional Sleep Disorders Center. It's like the "record" button in your brain has been switched off.

So why is it that some people regularly recall the intricate plots of their bizarre dreams? "The primary determinant of whether you're going to remember a dream is if you awaken during it," Mahowald says. Waking up snaps your memory into action, and it absorbs the bits of a dream that might otherwise fade into your subconscious. Frequent wakers are more likely to remember their dreams than deep sleepers, he adds. The content of your dreams might also play a role in retrieving your somnolent synapses. Bad dreams typically jolt you awake, so people who suffer from frequent nightmares may also be more likely to remember what they dream.

Research on dreaming has shown that it's possible to train yourself to better recall whatever you were dreaming about just prior to waking up, says Alan Manevitz, a psychiatrist at Weill Cornell Medical Center in New York City. "Keep a notepad by your bedside," he advises, "and in the first moments of waking, practice writing down whatever comes to mind." Unless, of course, you constantly dream of being chased by a certain undead serial killer named Freddy.

WHAT ARE SOME COOL THINGS TO DO WITH MY BODY AFTER I DIE?

GIVE SOMETHING BACK

If you like the idea of returning to the Earth after you depart, "promession" (a body-disposal method developed by Swedish biologist Susanne Wiigh-Masak) may be the way to go. A body is put in a container and dipped in a vat of liquid nitrogen cooled to −321°F (−196.1°C), which dehydrates it and makes it so brittle that a jolt of vibration "shatters" it into heaps of powder. After a machine removes dental fillings and artificial joints, the flour-like remains (equivalent to about a third of the body's original weight) are placed in a box made of potato starch. The shallowly buried container and its contents disintegrate in less than a year, returning essential nutrients such as nitrogen and phosphorus to the topsoil (coffins, in contrast, can take decades to decompose, and cremation turns corpses into nutrient-poor inorganic carbon). If the environmental implications don't convince you, consider the cost. Wiigh-Masak estimates that promession for one person will cost around $1,000—far less than the $8,000 that the average funeral and burial require. "There are six billion people on the Earth, and none of us is giving anything back to the soil," Wiigh-Masak says. "If I know my body can make a plant grow after I die, that's very appealing." The Swedish city of Jönköping hopes to convert its only crematorium into a "promatorium," and officials in the U.K., Germany, and South Africa are also showing interest in the procedure. Judging by that enthusiasm, Wiigh-Masak believes that the U.S. will soon follow suit.

THE FINAL FRONTIER

Our favorite alternative for the afterlife—being launched into space like James Doohan, *Star Trek*'s Scotty—is suitable for those not bothered by the vagaries of commercial space launches. A healthy bank account helps, too: Space Services Inc., the company that launched Doohan, charges $2,500 to put one gram of remains into Earth orbit. If you want to get to space a more cheaply, get your friends to hitch your remains to a weather balloon—you can pick one up for as little as $10, and it will get you to the edge of space. (Just make sure to check with the Federal Aviation Administration before letting it go.)

If you like the space launch idea because you want to go out in a big bang, you can try the Hunter S. Thompson route: Have your loved ones pack your ashes in fireworks and blast them from a cannon.

PUT A RING ON IT

Have your remains turned into a family heirloom. LifeGem in Elk Grove Village, Illinois, has perfected a flashy type of after-death alchemy: transforming cremated remains into man-made diamonds. Technicians extract pure carbon from the ashes, then place the carbon into a diamond press that exerts intense heat (about 2,200°F [1,200°C]) and pressures of up to 50,000 atmospheres for several days. The rough diamond that results—which will be naturally colored yellow or orange but can, through processing, be changed to blue, red, green, or colorless—can be faceted and polished just like an ordinary stone. In addition to choosing the color, you can specify the size of your LifeGem, from 0.2 to 1 carat. At $20,000 for a 1-carat blue diamond, though, it's a fair bit more expensive than a typical cremation, which costs around $1,000 (small yellow gemstones start at $2,700). Total turnaround time: six to nine months. LifeGems have attracted a devoted following of thousands who prefer their deceased loved ones around their fingers, not underfoot. "The diamonds are tangible, they're beautiful, and they can be handed down," says the company's CFO, Mike Herro. "Family members say we've made their difficult situation a little bit better."

095

ANY CHANCE
CROCODILES COULD LIVE IN THE ARCTIC?

> **SNACKING ON SEALS**

Fifty-five million years ago, the Canadian Arctic was as balmy as Florida and lousy with crocodile-like animals called champsosaurs. Could it happen again? The answer is yes, but probably not anytime soon. The Intergovernmental Panel on Climate Change estimates that the planet's average air temperature could warm by as much as 11.5°F (6.4°C) by the end of the twenty-first century. As a result, the world could be warmer than it was 55 million years ago, says Mark Lynas, author of *Six Degrees*, an analysis of hundreds of climate studies that reads like a nonfiction version of *The Day After Tomorrow*.

That said, the task of determining how individual species, much less entire ecosystems, will respond to rapid climate change is difficult at best. In the same regions where scientists found remains of champsosaurs, they also found fossils of their favorite food: turtles. Modern-day crocodiles could certainly be comfortable in a warmer north, but only if the prey and ecosystems required to support them proliferate there, too.

There's one last hurdle to overcome—even if the Arctic air warms up, there most likely still will be plenty of ice in the winters. The most aggressive climate models estimate that it will probably take thousands of years for the ice sheets to disappear year round, so cold-blooded crocs will have to wait at least that long to head to the poles.

CAN COCKROACHES REALLY SURVIVE A NUCLEAR HOLOCAUST? 096

THEY SEEM TO SURVIVE EVERYTHING ELSE

First of all, nothing would live through the intense heat at ground zero. For instance, the 15-kiloton bomb that exploded over Hiroshima ignited a 1,800°F (982.2°C) firestorm that incinerated everything within a radius of 1.2 miles (1.9 km). Outside that radius, however, a roach stands a pretty good chance of surviving the subsequent fallout.

The average cockroach can withstand a dose of about 6,400 rads (the standard measurement for radiation exposure). In comparison, the lethal dose for humans is only 500 rads—roughly the equivalent of 42 simultaneous full-body CT scans.

One theory on roaches' resilience credits their weekly larval molt, during which their cells divide half as frequently—and as adults, their cells divide even less often. Because radiation causes the most mutations in DNA that is replicating—which occurs most frequently in dividing cells—this slow replication protects roaches from radiation. So your kitchen's unpaid tenants may indeed be the ones building the next civilization after we check out.

097 ANY OTHER CONTENDERS FOR POST-HOLOCAUST OVERLORDS?

ROACHES VS. BACTERIA

When it comes to radiation resistance, roaches can't hold a candle to *Deinococcus radiodurans*, a nearly radiation-invincible bacteria. In 1956, scientists discovered the organism thriving in a can of spoiled meat that they had zapped with gamma radiation, and subsequent studies have shown that it can survive 1.5 million rads. That's enough, presumably, to live through the fallout from a 1.2-megaton bomb, which experts speculate is the largest in the U.S. arsenal.

Scientists are still at odds over just what DNA-repair mechanism allows *D. radiodurans* to withstand such extremes. One possibility is that right at the moment the bacteria's DNA strands break, a protein caps the ends and protects them long enough for the strands to reattach. To visualize this, imagine a torn rope's ends being immediately covered before they can start fraying uncontrollably. If this hypothesis turns out to be true, scientists might be able to introduce the same protein to human cells and enable us to outlast those pesky, indestructible cockroaches . . . giving us the last laugh after all!

HOW IS A DOLPHIN LIKE A RODEO BULL?

RIDE 'EM COW (OR DOLPHIN) BOY

For starters, both of them can do some pretty awesome acrobatics. But can we learn anything from studying these animals beyond that similarity? Well . . . kind of—if we look in their ears. All vertebrates have tiny structures called semicircular canals inside their ears that help them to maintain balance even when their heads are bobbing around vigorously. That's why rodeo bulls can buck wildly and not fall over.

The rule of thumb is that the greater the ratio of canal size to body weight, the more agile the beast. But that rule doesn't seem to hold for dolphins and whales: A bottlenose dolphin's canals are the same size as a mouse's. "Some people thought that, because there's nothing to bump into in the ocean, dolphins could get by with small canals," says Timothy Hullar, a (human) ear specialist at Washington University in St. Louis. "But dolphins are acrobatic swimmers—they can do barrel rolls and flips!—so there must be another explanation."

First, Hullar measured how much dolphins jerk their heads around by packing a sensor that records three-dimensional rotation in a waterproof tube and getting dolphins to swim around with it. "I didn't invent the sensor, but I figured out how to get a dolphin to keep it in his mouth," he says. "It took a lot of reward fish." He then analyzed the motions of a rodeo bull, which is just as active and similarly sized but has much larger canals than a dolphin. Hullar duct-taped the tube to a bull's horns and let it buck.

The device recorded similar head movements for bulls and dolphins, suggesting that dolphins' canals are either more efficient or that dolphins are able to maintain balance through some other mechanism. Hullar thinks dolphins and whales may receive more sensation through their skin than other animals, and that those sensations could supplement the cues from their canals. "If the animal feels water flowing over its body from left to right, maybe it can make the conclusion that it's turning left," he says.

Learning how dolphins keep their heads on straight, Hullar says, could help humans suffering from loss of function in the inner ear, a crippling affliction. "If dolphins can enhance their sense of balance with their sense of touch, we might be able to help these patients by stimulating another sensory input," he says. "It would be really neat if this is how we figured that out."

099

IS IT REALLY POSSIBLE TO SNEAK UP ON A SLEEPING COW AND TIP IT OVER?

▶ **AFTER-HOURS FIELDWORK**

We've all heard the story: A cousin's girlfriend's uncle's college roommate, sometime after last call on a Tuesday in Iowa, stumbles into a quiet cow pasture and ambushes a snoozing bovine, toppling it—thus becoming another initiate into the vast brotherhood of cow tippers. At the University of British Columbia, student Tracy Boechler had also heard the story, but she had her doubts. The animals weigh about 1,500 pounds (680 kg), she points out, so "actually tipping a cow is quite difficult when you take the biology and physics into consideration." Boechler took on the legendary "sport" for a zoological physics class project, emerging with evidence that claims of single-handed cow-tipping prowess are the result of overactive imaginations and too much alcohol.

BRING THE WHOLE TEAM

To tip the average cow, which is about 5 feet (1.5 m) tall, would require around 654 pounds (297 kg) of force, Boechler found. Assuming an athlete weighing 150 pounds (68 kg) can push her own body weight, the equivalent of 4.36 people would be needed to take down one cow. With some adjustments (such as binding the cow's legs together, to adjust weight distribution and the crucial pivot point), fewer people would be required, but that's also assuming the cow doesn't wake up and trot away. Although Boechler hasn't made it out into the field to test her calculations, she is confident about her conclusion: One drunken fool would be no match for a car-sized animal, no matter how top-heavy it seems. So the next time you're looking for post-libationary fun, you'd best bring along at least four buddies. Or better yet, why not just leave the beasts in peace?

IF A MOSQUITO BITES ME AFTER I'VE HAD A BEER, CAN IT GET DRUNK?

TEST THIS ON YOUR NEXT CAMPING TRIP

Shockingly, no major studies have been conducted on this topic. "The implications are, however, profound," says Michael Raupp, an entomologist at the University of Maryland. "Reckless flying, passing out in frosty beer mugs, hitting on flies instead of other mosquitoes. Frightening!" Fortunately, enough related research exists to make an educated guess.

The first thing we need to figure out is whether alcohol affects a mosquito's simple nervous system the way it does creatures with complex brains, such as dogs or Charlie Sheen. In labs, honeybees fly upside down after alcohol exposure, and inebriated fruit flies have trouble staying upright and (perhaps as a result!) fare poorly on learning tests. This suggests that mosquitoes, too, can get tipsy.

NOT CHEAP DATES

Now that we've got that settled, how much alcohol does it take to get them schnockered? Scientists routinely puff ethanol vapors at insects and measure their sensitivity with devices called inebriometers. Bugs are no lightweights, often withstanding vapor concentrations of 60 percent alcohol, far more than what's in our blood after a couple beers. "Someone who's had 10 drinks might have a blood alcohol content of 0.2 percent," says entomologist Coby Schal of North Carolina State University. To a mosquito, a blood meal that contains 0.2 percent alcohol is like drinking a beer diluted 25-fold.

SO I SHOULDN'T TRY TO PARTY WITH A MOSQUITO?

DON'T CHALLENGE IT TO BEER PONG

Indeed, skeeters can party hard, but they didn't learn how at college like the rest of us—scientists surmise that they might have developed this awesome ability to hold their liquor through their normal diet. In addition to your blood, they also feed on fermenting fruit and plants, which contain at least 1 percent alcohol and could boost their tolerance. And in a mosquito, alcohol (and any fluid other than blood) is diverted to a "holding pouch," where enzymes break it down before it hits the nervous system. In short, they're better equipped than humans are to handle alcohol.

So before you try to drink a mosquito under the table, heed this warning from Michael Reiskind, an entomologist at Oklahoma State University: The blood alcohol levels required to do so would almost certainly kill you as well.

DO SOME PEOPLE TASTE BETTER TO MOSQUITOES THAN OTHERS?

▶ SWEET, SWEET BLOOD

Ever notice that mosquitoes seem to be drawn to you like bees to a watermelon, while ignoring another person standing right nearby? There's some uncertainty about exactly what makes one person more appetizing than the next to a blood-hungry insect—but the "favorites" phenomenon is more than a suburban legend.

"The people who get bitten the most are the people producing the most carbon dioxide," says University of Florida mosquito authority Jonathan Day. But there's more to it than CO_2. "If they were just attracted to carbon dioxide, they would go for cars," Day says. Perfume, body heat, and exercise-induced lactic acid can all turn a mosquito on to her meal. (Only female mosquitoes drink blood. Males get by on fruit juice and nectar.)

The sex hormone estradiol is also a mosquito magnet. Although women have more estradiol than men, any bite resistance guys have is canceled out by their higher body mass, which produces more CO_2. The best way to avoid being dinner is to ditch the cologne and keep the aerobics to a minimum.

WHY ARE GLOW-IN-THE-DARK ANIMALS USEFUL TO SCIENCE?

Scientists have created a series of glowing lab animals in recent years, from pigs to naked mole rats to an extremely adorable beagle named Ruppy the Glowing Puppy. Why, you might very reasonably ask? Aside from the obvious benefits of being able to find your cat in the dark, splicing in the jellyfish gene that causes these animals to glow under UV light is a simple way to test methods of genetic engineering. Did your painstaking transgenic experiment work? Just turn on the blacklight and find out.

104

WHY WOULD YOU EVER, EVER CROSS A GOAT WITH A SPIDER?

The spider-goat, which luckily does not have eight extra-giant spider legs, is a surprisingly useful hybrid creation. Randy Lewis's team at the University of Wyoming has engineered these mostly goatlike creatures to produce milk that can be dried, purified, and spun into spider silk—a half-ounce (14 g) of silk from 1 quart (.9 liters) of milk, which equals the output of a hundred spiders. Spider silk is extremely strong and light, and theoretically usable in human endeavors from bioengineering to fashion. If the spider-goat catches on, we may finally be able to mass-produce this fabulous material.

105

IS IT SAFE TO EAT GENETICALLY MODIFIED FISH?

Your lox will never be the same. AquAdvantage salmon (from AquaBounty Technologies) have been bred to blend the properties of three fish: Atlantic salmon, Chinook salmon, and the colorfully named ocean pout. The genetic combination creates a fish that produces growth hormone constantly, allowing it to grow very large, very quickly. The FDA has pronounced the fish safe, although consumer protection groups including Consumer Reports have voiced concerns over the supersized animals, calling the FDA's report incomplete. Even though the idea of eating giant Frankenfish is a little weird, we tend to side with the FDA on this one.

CAN WE MAKE FACTORY FARMING KINDER?

To combat the cruelty of factory farming, some scientists have suggested genetically modifying beef cattle to dull or eliminate their ability to feel pain. A paper in the journal *Neuroethics* suggested this somewhat extreme solution to the abuses endured by these animals—rare genetic abnormalities in humans can produce people unable to feel pain normally, and it could be possible to engineer a similar deficiency in cattle. While the cow's life wouldn't be much richer, it could at least be more painless.

HOW CAN MOUSE MILK IMPROVE BABY FORMULA?

Baby formula may work just fine, but it doesn't contain all the organic properties of human breast milk. A team of Russian scientists is trying to replicate some of these missing elements by splicing human genes into mice. Hopefully, this will allow the mice to produce lactoferrin, a protein found in breast milk (but not formula) that provides protection to human newborns. If the scientists succeed, they can use the technique to produce the protein in larger animals like goats that aren't quite so tricky to milk, potentially bringing formula a lot closer to the real thing.

108
WHAT DO ANIMALS DREAM ABOUT?

▶ INSTANT REPLAY

When Spot pedals his legs in a futile sprint to nowhere during a nap, he's probably reliving the morning's game of fetch or his latest attempt to catch a squirrel. Scientists have found that, as with humans, what goes on in animals' brains while they're sleeping is influenced by what they did that day.

How do we know this? As with so many scientific discoveries, we have rats to thank. In his lab at Massachusetts Institute of Technology, neuroscientist Matthew Wilson and his colleagues implanted electrodes in rats' brains to record just what goes on

in there while the critters are catching some shut-eye. Wilson observed the rodents' brain activity first as they ran through a maze and then when they were sleeping afterward. The activity in a sleeping rat's hippocampus—the area most responsible for forming autobiographical memories—matched patterns detected while they had been in the maze.

At the same time, the visual cortex replayed corresponding sequences, too, suggesting that the sleeping rats were remembering not only how they scurried through the maze, but also what they had seen—all in the order that it had actually occurred.

109   DO RATS DREAM ABOUT ANYTHING OTHER THAN MAZES?

▶ THEY WOULD IF WE EVER LET THEM OUT

Of course, rats dream about their time outside of the maze, too. Matthew Wilson's team has recorded many unidentified brain-activity patterns, which they hypothesize are memories of the rodents' "free time" spent sitting in their cages or hanging out with other rats. The simple daily life of rats and other animals makes for what appear to be unimaginative dreams. "People's dreams tend to be more complicated, with bizarre content," Wilson says, "because our life experience is more complex than many animals'."

Replaying the day's events, he says, might be a way for animals and humans to learn from the past in order to make better choices in the future. For example, dreaming about the maze might help a rat take a more direct route to the cheese on its next attempt. "That," Wilson says, "is the kind of learning you don't have time for when you're awake."

110
IS IT TRUE THAT A
QUACK HAS NO ECHO?

▶ **ONLY FOR NINJA DUCKS**

For reasons most likely involving rumor and mass gullibility, the myth that a duck's quack has no echo has gained a surprising amount of traction over the years—enough traction, in fact, to prompt an acoustical research team at the University of Salford in England to undertake the conclusive study and debunk the legend.

The team recorded Daisy the duck in a cathedral-like chamber designed for maximal echo formation and then, as a scientific control to isolate the quack with no echo, in an echoless room with walls covered in fiberglass wedges that absorb all sound. The result: Daisy's quack did indeed echo in the first room.

So why the misguided belief? A duck's quack, like any sound, travels in waves. When the waves hit a reflective surface, they bounce back, and the echo is the expression of that reverberation.

If, however, the beginning of the echo reaches your ears at the same time as the original sound is ending, the sounds will naturally overlap, becoming indistinguishable. In the duck's case, the quack begins softly and ends loudly, so the soft early echo is sometimes drowned out by the louder original end sound (the "aack!" part). Trevor Cox, who led the Salford team, speculates that this high/low dichotomy is what originally provoked the anti-echo belief.

In fairness, the Salford study isn't the first. A few years ago, a team from the newspaper column "The Straight Dope" tackled the same myth out in the wild, though with slightly less reliance on the scientific method. The researchers simply coaxed a duck to quack (by running with the bird in their arms) in a high school courtyard and listened for an echo—which, they reported, returned quite clearly.

111

WHY DO **DUCKS** HAVE ORANGE FEET?

▶ **CHECK OUT THOSE LEGS**

Actually, many species of ducks have feet and legs tinted a bluish green or gray. But for the ducks that do strut around on orange feet, well, it's all about attracting the ladies. Chicks dig orange.

Kevin Omland is an evolutionary biologist at the University of Maryland at Baltimore County, and he knows as much about mallard-duck coloring patterns as anyone; it was the topic of his graduate thesis. "I looked at male mallards and thought, gosh, they exhibit so many wonderful colors; I wonder which ones females care about," he says. Do lady ducks lust after the males' green head plumage? Or maybe it's the blue patches on the males' wings? Then again, what female duck can resist a nicely proportioned set of white "neck tie" feathers? After four years of documenting mallard courtships, Omland found that none of those features mattered. All the female ducks cared about was the brightness of the guy's yellow-orange bill.

Bright orange coloring suggests that a male duck, also known as a drake, is getting all his vitamins, particularly carotenoids, such as beta-carotene and vitamin A, which are antioxidants that can be beneficial to the immune system. "This indicates that his behaviors and genes are good enough for him to recognize and eat the right food, or that his immune system is strong enough to produce bright orange legs," Omland says. "The female sees this as a very attractive trait to pass on to her offspring."

Omland's work only looked at drakes' bills, but he thinks there's enough circumstantial evidence to say that ducks check out each other's feet, too. "Blue-footed boobies have, obviously, very blue feet, and it's very well documented that they use their feet in courtship and that females do care about the coloration of males' feet," Omland says. "Perhaps mallards, like the boobies, have a foot fetish."

IF **EVOLUTION** HAS TAKEN A DIFFERENT TURN, COULD
DRAGONS HAVE EXISTED?

112

▶ LET THE CROSSBREEDING BEGIN

It would have taken quite a few turns for natural selection to have produced dragons, but if you're willing to stretch your criteria a bit, most classic dragon characteristics do exist in other species. They just don't come packaged in one animal.

First up on the dragon checklist is flying. Dragon wings are usually depicted in one of two ways—a third pair of limbs connected to the backbone, or a pair of webbed forearms. Jack Conrad, a paleontologist and reptile expert at the American Museum of Natural History in New York, thinks the latter is more plausible. "It seems that six appendages are very unlikely in vertebrates," he says.

"The only thing close to having six limbs are these frogs in the western part of the U.S. that get this bad parasite and end up generating extra limbs. Even then, the new limbs are identical to the hind limbs, and the frogs don't do well. It seems that anytime nature tries to generate a vertebrate hexapod, it dies. That seems to be the main limitation."

In Conrad's opinion, the leathery wings of a pterosaur are the best possible flight mechanism for your basic giant lizard. "*Quetzalcoatlus* had a wingspan of 30 feet [9.1 m]," he says. "That would do the trick."

But strong wings would be necessary to compensate for the weight of a dragon's skin, which, of course, would need to be able to repel bow-and-arrow attacks. "Let's throw a little alligator in there for armor," Conrad says. An alligator's skin, he explains, is made partly of bony plates. When European settlers first encountered the reptiles, the skin proved to be tough enough to turn away a musket ball, plenty strong for a dragon.

OK, so we've got a very large alligator with the wings of a pterosaur that can withstand musket fire. Now it just needs to breathe flames. This is where no parallel exists—there are no known animals that can spit fire or even a flammable liquid. But there are some beetles that can shoot caustic chemicals from their abdomens that can burn people's skin, so it's not totally out of the question that some animal at some point in time could make a flammable liquid.

Add to this the fact that cobras can spit venom accurately at objects 6 feet (1.8 m) away, and we can imagine our dragon mimicking that ability to propel the flammable liquid. As for lighting it? "Well, maybe, if you had some specialized organ like an electric eel's tail dangling in the mouth, that could spark that liquid and allow the creature to breathe fire," Conrad says. "Of course, this is all very theoretical."

113
IS IT TRUE THAT
BIRDS CAN'T FART?

▶ **THEY'RE TOO CLASSY**

It's not that they can't. They just don't need to, says Mike Murray, a veterinarian at the Monterey Bay Aquarium in California. Birds have the anatomical and physical ability to pass gas, he explains, "but if I saw gas in a bird's gastrointestinal tract on an X-ray, I'd suspect that something abnormal was going on in there."

Birds don't typically carry the same kinds of gas-forming bacteria in their gut as humans and other mammals do to help digest food, so there's nothing to let loose. Parrots sometimes emit fartlike sounds, but it's not what you might think. "They like to make playful sounds like they're giving you a raspberry, but it's coming from the north end, not the south," Murray says.

BUT CAN THEY BURP THE ALPHABET?

Scientists are a little less certain about whether birds can release gassy buildup from the mouth, though. There's no official documentation of a bird burp (it's not a common field of research), but most ornithologists suspect that if a bird needed to burp, it would have no trouble doing so. "Birds are able to excrete lots of things through their mouths," says Todd Katzner, the director of conservation and field research at the National Aviary in Pittsburgh. "The fact that birds can regurgitate food for their young suggests that they can also reverse the direction of other things down there. I'd be pretty surprised if birds didn't burp."

WHY DON'T I EVER SEE BABY PIGEONS? 114

▶ **THEY SPRING FULL-GROWN FROM THE EAVES**

Run-of-the-mill city pigeons, known as rock doves, build their nests in the nooks and crannies of the concrete cityscape, which are reminiscent of their native European and Middle Eastern cliffside habitats. Parents typically keep their babies, or squabs, hidden and safe until they can survive on their own, usually a month after they hatch.

As a result, youngsters are almost fully grown and their feather coloring looks nearly identical to an adult's by the time they fly the coop, says Karen Purcell, who leads Cornell University's Project PigeonWatch, a grassroots study of feather colorings.

115 SO HOW DO I SPOT A TEENAGE PIGEON THEN?

▶ **BY THEIR BAD ATTITUDE**

Karen Purcell suggests a few tricks for spotting the younger birds among the masses. Pigeons have grayish-brown eyes for the first six months of their lives, after which they turn orange or red. And the bit of flesh above a pigeon's beak, called the cere, is gray when the bird is younger, instead of white. You can also identify juveniles by their behavior, Purcell says—although it's hard to tell which birds are acting immature when all of them are pooping on statues.

116 IS IT TRUE THAT DOGS SWEAT THROUGH THEIR TONGUES?

SOMETHING TASTES ODD

During the dog days of summer, a slobbery tongue cools a dog just as well as your sweaty armpits do for you. But fortunately for dogs, their tongues don't actually produce sweat.

Animals with sparse body hair—including humans, horses, and some species of monkey—cool down as sweat evaporates off their bodies. For long-haired species, such as dogs, sweating would just result in a soggy coat. Instead, dogs stick out their tongues and pant to chill out.

Energy, in the form of body heat, is required in order to evaporate liquid off the surface of the skin or tongue, explains Jack Boulant, a thermal physiologist at Ohio State University. As this heat vaporizes surface moisture, body temperature drops.

Over the years, scientists have learned that a dog's internal thermostat, the thermoregulatory system, responds to heat by pumping hot blood to the tongue, opening the saliva floodgates, and triggering fast, shallow breathing. As the warm air flows over the wet respiratory tract and tongue, it helps evaporate the moisture, sending heat away from the dog's blood.

In addition to lowering body temperature, this process helps to refrigerate the brain. The blood drains from the nose and tongue and chills the blood flowing to the brain, which keeps the heat-sensitive organ at a lower temperature than the rest of the body. The cooling system works less well for short-muzzled breeds, such as Pekingese, which have smaller noses and squished air passages.

117 ANY OTHER COOL COOLING SYSTEMS IN THE ANIMAL WORLD?

AVOID STORKS ON HOT DAYS

In fact, dogs aren't the only animals that use clever cooling tricks. Rats lick their bellies. Resting kangaroos pant and lick their bodies, and hopping makes them break a sweat. And this may not sound cool, but storks find extra relief by defecating on their own long legs.

WATER TORTURE

A misbehaving cat caught by a spray bottle doesn't necessarily run away because it hates getting soaked. It's probably afraid of the actual water. Scientists believe that house cats' limited experience with water—mostly leaky faucets and water dishes—makes them fearful of wet stuff.

This is likely because the domestic cat's desert-dwelling ancestors also had limited experience with water. Genetic studies conducted at the National Cancer Institute show that the house cat's closest relatives are wild cats from Africa and Europe and the Chinese desert cat. And ever since humans first domesticated cats—the earliest evidence dates to Cyprus 9,500 years ago—cat owners have protected their pets from the harshness of the elements. "Cats have not evolved to do much with water," says animal-behavior specialist Katherine Houpt of Cornell University's College of Veterinary Medicine.

Whether a cat enjoys water depends not only on where it lives, but also on its interactions with predators and prey, says Jack Grisham, the St. Louis Zoo's director of animal collections. Lions stay on dry land to avoid bathing in rivers patrolled by crocodiles, and leopards live in trees, away from water and predators below. In contrast, some domesticated farm cats prowl ponds hunting for frogs. And the fishing cat, native to wetlands from India to Indonesia, taps the water's surface with its paw and then snatches its prey with webbed claws.

Owners can also train the fear of water out of cats by bathing them as kittens. Most vets don't recommend this, however, because it can dry out a cat's skin and wash away pheromones essential for communicating with other felines. Besides, a cat already has all the supplies it needs to keep itself clean: Its saliva contains a natural detergent to reduce grease, and its barbed tongue combs out dirt.

Still, some cats actually enjoy getting wet—so much so, Houpt says, that "they actually play with water." Even the threat of the spray bottle, she admits, stops only 70 percent of cats from scratching the sofa.

119
DO CLONED WILD ANIMALS HAVE INSTINCTS?

▶ **BORN TO BE WILD**

If you raised a cloned tiger as a sweet little pussy cat, would it curl up on your lap? Or is there feral behavior coded in its DNA, just waiting to kick in? Let's ask Betsy Dresser, former senior vice president of research at the Audubon Center for Research of Endangered Species in New Orleans, who has raised several litters of small African wildcat clones. "Oh yes, the clones are very much wild animals with wild instincts," she says. "They bite and scratch. You can't handle them without gloves and nets."

Dresser uses domestic cats as surrogate mothers for cloned African wildcat embryos, and although a tabby mother can calm the kittens, her influence doesn't last. "They aren't as hissy, and they don't fight as much," she says. "But once you get them away from domestic cats, especially once puberty sets in, their aggressive survival behavior emerges."

Clones aren't blank slates, Dresser explains. They're exact genetic copies of another creature. The behaviors that make African wildcats successful hunters in the savannah are, fundamentally, made possible by the activation of just the right gene at just the right time. The first African wildcat whose DNA told its brain, "Hey, eat that field mouse," stood a better chance of surviving and reproducing, and when it did, its offspring inherited that trait and automatically expressed the same survival behavior. "Those genes pass on when you clone an animal, too," Dresser says. "I think our clones' behavior makes a strong case that instincts are at least partly genetic."

So if scientists ever clone a saber-toothed tiger, it unfortunately won't end up in a Las Vegas magic act—it would probably rip your arm off. And sadly, a resurrected dodo wouldn't know how to avoid repeating history. It would just stand around placidly like they all did, waiting to be clubbed back into extinction.

CAN WE CLONE
EXTINCT ANIMALS?

PREHISTORIC RERUN

It's looking more and more likely that scientists will be able to resurrect some lost members of the animal kingdom through cloning. Disappointingly, dinosaurs would not be first on the list—more recently vanished species would offer the most viable DNA samples for reconstruction.

A Japanese team led by Akira Iritani, professor emeritus of Kyoto University, is hoping to deliver a real, live woolly mammoth within five or six years. Mammoths are unusually good candidates for resurrection: Although they've been extinct for thousands of years, their northerly habitat means that numerous mammoth bodies have been found entombed in ice. Although freezing damages DNA, Teruhiko Wakayama of the Riken Center for Developmental Biology has developed a technique for salvaging viable DNA from long-frozen mice that the mammoth team has adapted to extract undamaged nuclei from mammoth egg cells.

There's a lot of work still to do, however. The mammoth egg nuclei will need to be implanted in elephant egg cells, and the (hopefully) viable embryo that results would then need to be carried to term by an elephant mother—a process that may well present new problems, despite the strong genetic similarity between mammoths and elephants. But with a little luck and a lot of scientist-hours, we may have our very own baby mammoth to study. And from there, who knows? Pet dinosaurs could be closer than we think.

WHAT IS THE SMALLEST ANIMAL IN THE WORLD?

▶ **. . . AND IS IT CUTE?**

It's complicated. The smallest animal is a species of crustacean, but when most people think "animal," they generally mean "vertebrate," so that's what we'll focus on. As it turns out, scientists around the world are arguing over who has the smallest one—and all the contenders are fish.

The feud started when a team of researchers headed by biologist Maurice Kottelat of the National University of Singapore said that it had found the world's smallest vertebrate. The tiny creature was a miniature carp living in acidic peat swamp forests on the Indonesian island of Sumatra. The transparent *Paedocypris progenetica* is less than a third of an inch (8 mm) long and lacks the top of its skull.

But in their report on the new species, the scientists overlooked an earlier discovery by biologist Ted Pietsch of the University of Washington. A paper Pietsch published in September 2005 in *Ichthyological Research* describes a male anglerfish measuring a mere quarter of an inch (6 mm). Collected in deep water in the Philippine Sea, *Photocorynus spiniceps* attaches itself to its mate for life by biting onto her side, back, or belly.

Pietsch's fish may be the smallest on record, say his foreign rivals, but it doesn't qualify as the smallest fish species because the female anglerfish is more than seven times the size (measured from tip to tail) of its mate. The male, which has huge eyes and testes, is basically just a parasite. (Pietsch is quick to point out that the male anglerfish is not a total degenerate—it does have other body parts; they're just less prominent.)

A third team of scientists who described yet another species of tiny fish in 2004 wonder why their colleagues are so hung up on length. H. J. Walker of the Scripps Institution of Oceanography and William Watson of the Southwest Fisheries Science Center, both in La Jolla, California, point out that even if *Schindleria brevipinguis*—a Great Barrier Reef inhabitant better known as the stout infantfish—isn't shorter than the other contenders, males and females alike are so slender that some weigh as little as 0.7 milligrams. That's less than a mosquito—and either of the other two species.

One may wonder why scientists care at all. In fact, they hope to understand the limits of vertebrate physiology and why some fish evolve to become so tiny. One theory is that small size helps the fish survive in nutrient-poor habitats such as peat swamps and the deep ocean. Another theory posits that small size helps them evade predators.

As for the other end of the scale, there's no contest. The blue whale, which can grow to a length of more than 100 feet (30 m) and weigh upward of 150 tons (136,000 kg), is the biggest.

122 COULD A HUMAN BEAT A T. REX IN AN ARM-WRESTLING MATCH?

HE BETS YOU FIVE BUCKS YOU CAN'T

"First, we're assuming that the T. rex won't just eat the person, right?" asks Jack Conrad, a vertebrate paleontologist at the American Museum of Natural History in New York. Right. This is a sanctioned match, and killing your opponent is strictly against the rules. "Doesn't matter," Conrad says. "There's no chance that any human alive could win."

The T. rex's arms might look wimpy, but they were extremely strong. Each was about 3 feet (0.9 m) long, and based on the size of the arm bones and analysis of the spots where muscle attached to the bone, they were jacked. "The bicep alone—and this is a conservative estimate—could curl 430 pounds [195 kg]," Conrad says. Even the beefiest humans max out at around an embarrassing 260 pounds (117.9 kg). Surely an *Over the Top*–era Sylvester Stallone would put up a good fight? "Not even Lou Ferrigno in his prime would stand a chance," Conrad says. "T. rex didn't just have big biceps. Their chest and shoulder muscles were huge, too. They had huge arms and shoulders—bigger than my leg. They had the strength to rip a human's arm right out of its socket."

123 REALLY? THERE'S NO WAY I COULD WIN?

MAYBE BY FORFEIT

Humans, comparatively puny as they might be, could still have one advantage. There is actually a chance that your competition might not be able to put his beefy muscles to use. That's because we're not totally sure exactly how he used all that muscle. There are dozens of hypotheses about how T. rex used its arms, Jack Conrad explains, but the ones taken most seriously involve pushing itself up if it was lying on its belly, tossing big chunks of meat into its mouth, or holding onto females during what scientists suspect was a very vigorous mating routine. These ideas are favored because such actions required Barbie doll–like up-and-down motions of the arm, and fossil evidence indicates that the dino king was incapable of rotating or twisting its arms. "The T. rex probably couldn't have done the arm-wrestling move," Conrad says. "So maybe you could get him on a technicality."

124
HOW LONG WOULD IT TAKE FOR PIRANHAS TO EAT A PERSON?

▶ **LET'S NOT TEST THIS ONE OURSELVES**

After a trip to the Amazon jungle, President Teddy Roosevelt famously reported seeing a pack of piranhas devour a cow in a few minutes. It must have been a very large school of fish—or a very small cow. According to Ray Owczarzak, assistant curator of fishes at the National Aquarium in Baltimore, it would probably take 300 to 500 piranhas five minutes to strip the flesh off a human weighing 180 pounds (81.6 kg). But would this attack even happen?

The thing is, piranhas get a bad rap. Yes, they are carnivorous critters with sharp teeth. "It's like they have a mouthful of scalpels," says Erica Clayton, Amazon collection manager at the Shedd Aquarium in Chicago. Even so, instances of piranhas biting humans are extremely rare. Most are happy snacking on other fish and plants.

In general, if you leave the toothy fish alone, they'll do the same for you. Still, if you decide you must take a dip in the Amazon, make sure you don't have any open wounds—the smell of blood attracts piranhas.

125 WHAT DO WHALES SING ABOUT?

▶ SINGING THE BLUES

For some insight into the world's largest singers, we asked David Rothenberg, a professor of philosophy and music at the New Jersey Institute of Technology whose book, *Thousand Mile Song*, analyzes whale songs.

"Humpback whales sing some of the most beautiful songs in the animal world. It's not just 'woo, woo, woo'—their songs last 10 to 15 minutes and have a definite form, usually consisting of five or six unique phrases. Only the males sing, which has led many scientists to theorize that they croon to attract females. The hole in this argument, though, is that no one has ever actually seen a female whale show any interest at all in a male's song.

"Male songbirds change their tunes to impress potential mates, but a group of male humpbacks all sing the same song. If the song changes midseason, they all adopt the same change. We don't really know why they sing together. They might be trying to create a sense of peace before they mate, or they could be staking out their territory. Either way, it makes the competitive-mating theory seem less believable.

"We're also not quite sure why they change their songs in the first place. It could be that one whale tweaks part of the song, and if it's catchy, the rest pick it up quickly. I gave this a shot by playing my clarinet to a whale swimming under my boat, and he seemed to change his song in response.

"Another theory is that whales' brains are programmed to change the tunes no matter where they are in relation to other whales. For example, scientists have made recordings of humpbacks in Hawaii and the Gulf of Mexico altering their songs in similar ways at the same point in the mating season, even though there's no way the groups could be hearing each other's songs.

"Most of the research money goes to studying whale songs for conservation efforts (each whale has a unique voice, so it's a good way of estimating how many are out there), not translating their meaning. But this much seems clear: If it takes 15 minutes to sing it, the message probably isn't too urgent."

LET'S HOPE NOT

"It's not as silly a question as you might think," says Michael Moore, a marine-mammal research specialist at Woods Hole Oceanographic Institution in Massachusetts. "It would take some extraordinary circumstances, but any mammal can get rabies."

Bats, coyotes, foxes, and raccoons are the most common carriers of rabies, but being landlubbers, it's highly improbable that any of them would have a chance to bite and infect a whale. One of those animals could, however, bite a seal that's resting on a beach, and then that seal could swim off and bite a whale. Although there is absolutely no record of a rabid whale, and only one documented case of rabies in a seal—a ringed seal caught in 1980 in Svalbard, an archipelago off Norway—the scenario may soon be of greater concern. "Starting 10 years ago, coyotes began to prey on harp seals here on Cape Cod," Moore says. "Because of that, I like for my staff to get vaccinated. There's a very small chance that a seal will have rabies."

Seals aren't known to attack whales (it's a size thing), but rabid animals behave erratically, so it could happen. Even if a rabid seal did bite a whale, it might take years for the whale to show symptoms. To become infected, the virus must travel along a nerve from the bite location to the central nervous system and brain. This is why a person bitten in the face by a rabid fox will show symptoms earlier than if that person had been bitten in the foot. Rabies travels along nerves at a rate of 0.3 to 0.8 inches (8–20 mm) a day, so if a whale measuring 50 feet (15.2 m) was bitten in the tail, it might take two to five years for the virus to reach the animal's brain and manifest.

127 SO WHAT SHOULD I LOOK FOR?

DRUNK-LOOKING WHALES

How would you ever even identify a rabid whale? "Well, the telltale foamy mouth would be very difficult to spot in the water," says Gregory Bossart, the chief veterinary officer at the Georgia Aquarium in Atlanta. "But as with other animals, rabies would interfere with any activity that involves the central nervous system, so a whale might exhibit abnormal swimming patterns or lose the ability to swim altogether. It might also have trouble with echolocation." Watch out, then, for zigzagging whales bumping into stuff. Another classic symptom of rabies infection is hydrophobia, which would be quite difficult for a whale to deal with. "Who knows?" Moore jokes. "Perhaps that's why whales strand themselves on beaches."

128
DO TORNADOES EVER HIT BIG CITIES?

▶ **MIDTOWN MAELSTROM**

It's true that the wide-open plains of Kansas are a more familiar backdrop for tornadoes than Times Square, but the funnels can form just about anywhere if the conditions are right.

The reason Tornado Alley, the area stretching from Texas to South Dakota and from the Rocky Mountains to Kansas, is the most active tornado spot in the U.S.—it sees hundreds a year—is not because it's flat farmland. It's because tornadoes form when two opposite weather systems collide under certain conditions, and this occurs with great regularity in Tornado Alley.

During springtime in that region, a constant stream of cool, dry air blowing southeast from Canada runs into a similarly steady stream of warm, moist air moving northwest from the Gulf of Mexico. As these weather fronts interact, they build high-intensity thunderstorms that, if they're strong enough, can create a powerful updraft of air. Low pressure at ground level and in the middle or upper atmosphere interacts with the rising air to create a swirling vortex that can eventually extend a tornado funnel to the ground.

129 CAN SKYSCRAPERS PREVENT TORNADOES?

▶ **I DON'T THINK WE'RE IN KANSAS ANYMORE**

It just so happens that most cities with a lot of skyscrapers are situated in places where tornado-feeding conditions evolve less frequently. But tornadoes do sometimes hit cities, says Gary Conte, a warning coordination meteorologist at the Upton, New York, outpost of the National Weather Service, citing recent touchdowns in Dallas, Memphis, Miami, and four of New York City's five boroughs (Manhattan has been spared, so far). Skyscrapers and topography don't matter. "Tornadoes form thousands of feet above building tops," Conte says. "Skyscrapers won't prevent the funnel from coming down, but they might influence its shape so that it doesn't look as nice and neat as it does on a flat surface like the plains. That doesn't make it any less of a tornado, though."

130

ARE WE LIVING IN A COMPUTER SIMULATION?

▶ **TAKE THE RED PILL**

Ever wonder if you're just a line of code in some giant, superpowerful computer? Sure, we all do. But Nick Bostrom, a transhumanist philosopher at the Future of Humanity Institute at Oxford University, has thought this possibility out a little more thoroughly than most of us.

"The notion that we exist solely in someone else's computer simulation isn't as sci-fi as you might think. It wouldn't be like in *The Matrix*, in which unconscious people were plugged into the simulation. Instead, the most likely alternate-reality scenario is that we're bits of information in an algorithm running on an incredibly powerful computer built by a superadvanced post-human civilization that is interested in learning about how their ancestors lived and evolved as a society.

"I arrived at this idea by positing that one of the following hypotheses has to be true. One: Humans will become extinct before they're able to build computers powerful enough to run simulations capable of creating entire virtual worlds full of people with virtual intelligence. Two: An advanced post-human civilization will have these computers but no desire to run simulations of their ancestors. Or three: We're all already living in the simulation, and this page—like you—is just a series of 0s and 1s.

"These are the only options, and because there's no real reason to choose one over the other, each has a probability of one-third. My gut feeling is that, in fact, there's just a 20 percent chance that we're living in a simulation. That hunch, however, inflates the chances of options one and two, which is a bit upsetting because I'd like to think that we as humans will not go extinct—and that we're interesting enough to warrant complex simulation."

131
WHAT HAPPENS WHEN WE
HALLUCINATE?

▶ **WHOA, CHECK OUT MY HAND**

Although most of us think of a *Fear and Loathing in Las Vegas* experience, or perhaps that weird guy talking to himself on the street corner, recreational drugs and psychiatric disorders are not the only causes of hallucination. Stress, fever, illness, and sleep deprivation also can trigger an episode.

Hallucination, also called sensory deception, happens when a person sees, hears, feels, smells, or tastes something that is not there. The general cause is abnormal chemical reactions, triggered by a drug or by misfiring neurons that activate certain parts of the brain and disrupt their usual functions.

The exact nature of hallucinations is poorly understood, but here's what we do know: With visual hallucinations, foreign chemicals (drugs or stray neurotransmitters) enter the synapses between the optic nerve and the occipital lobe, the part of the brain that processes visual information, triggering a signal on that neural pathway. Once the false signal reaches the brain, the occipital lobe is activated, and visual hallucination occurs. The same process occurs with hallucinations related to hearing, smell, taste (in the temporal lobes), and touch (in the parietal lobe).

If you wish to experiment (legally, mind you) with some hallucination of your own, try sleep deprivation. According to Michael Golder, a professor of psychiatry at George Washington University, "A person who has been sleep-deprived for 72 hours is as susceptible to hallucinations as someone taking LSD."

IF THE SUN WENT OUT, HOW LONG WOULD LIFE ON EARTH SURVIVE?

NO WORRIES, WE HAVE A SUN LAMP

Well, let's put it this way: If you put a steamy cup of coffee in the refrigerator, it wouldn't immediately turn cold. Likewise, if the sun simply "turned off" (which is actually physically impossible), the Earth would stay warm—at least compared with the space surrounding it— for a few million years. But we surface dwellers would feel the chill much sooner than that.

Within a week, the average global surface temperature would drop below 0°F (–17.8°C). In a year, it would dip to –100°F (–129°C). The top layers of the oceans would freeze over, but in an apocalyptic irony, that ice would insulate the deep water below and prevent the oceans from freezing solid for hundreds of thousands of years. Millions of years after that, our planet would reach spitting distance from absolute zero, a stable –400°F (–240°C). At that temperature, the heat radiating from the planet's core would equal the heat that the Earth radiates into space, explains David Stevenson, a professor of planetary science at the California Institute of Technology.

WE'LL ALL LIVE IN A YELLOW SUBMARINE

Although some microorganisms living in the Earth's crust would survive, the majority of life would enjoy only a brief post-sun existence. In the darkness, photosynthesis would halt immediately, and most plants would die in a few weeks. Large trees, however, could survive for several decades, thanks to their slow metabolism and substantial sugar stores. With the food chain's bottom tier knocked out, most animals would die off quickly, but scavengers picking over the dead remains could last until the cold killed them.

Humans could live in submarines in the deepest and warmest parts of the ocean, but a more attractive option might be nuclear- or geothermal-powered habitats. One good place to camp out: Iceland. The island nation already heats 87 percent of its homes using geothermal energy, and, says astronomy professor Eric Blackman of the University of Rochester, people could continue harnessing volcanic heat for hundreds of years.

Of course, the sun doesn't merely heat the Earth; it also keeps the planet in orbit. If its mass suddenly disappeared (this is equally impossible, by the way—after all, where would it go?), the planet would fly off, like a ball swung on a string and suddenly released.

133

HOW LARGE WOULD AN ASTEROID NEED TO BE TO SPLASH ALL THE WATER OUT OF THE PACIFIC OCEAN?

▶ **GRAB YOUR SURFBOARDS**

First of all, if a giant space rock were to hit the ocean, it wouldn't splash at all. The heat that a city-sized asteroid would generate passing through the atmosphere and the kinetic energy released on impact would be so enormous that the Pacific would immediately evaporate, says Lindley N. Johnson, the program executive for NASA's Near Earth Object Program.

But for the sake of argument, we'll ignore heat and consider the scenario to be roughly equivalent to dropping a bowling ball into a bucket full of water. Because the bowling ball's mass and density are greater than water's, Johnson explains, it splashes nearly all the water from the bucket. Similarly, to splash all the water from the Pacific, an asteroid would need to have a mass equal to, or greater than, the water it displaced.

Using the estimated volume of the Pacific Ocean (161 million cubic miles/671 million km³), and assuming the asteroid in question is spherical and has the same density as most space rocks (two to three times that of water), Johnson calculates that an asteroid 675 miles (1,086 km) in diameter, about one-third the size of the moon, would do the job.

The good news? "We don't know of any asteroids that large," Johnson says. The largest known asteroid is Ceres, which is 590 miles (949.5 km) in diameter. But it's anchored in the asteroid belt between Mars and Jupiter. In fact, such collisions are extremely rare: No object of that size has hit Earth since the impact that created the moon 4.5 billion years ago.

But that doesn't mean the news is all good. There's no real need to do that much splashing—a space rock one-tenth the size of Ceres would still generate enough heat to evaporate the entire ocean. And, Johnson points out, the asteroid that killed off the dinosaurs was just 6 or 7 miles (9 to 11 km) in diameter: "Getting hit by something even one-millionth of the size we're talking about would be a pretty bad day."

134

JUST HOW OLD IS DIRT?

REALLY, REALLY OLD

"It depends on what you mean by dirt," says Milan Pavich, a research geologist with the U.S. Geological Survey. "The oldest sedimentary rocks are about 3.9 billion years old—they're in Greenland—and at one time, they were dirt. That's pretty close to the time the Earth formed."

But those rocks are just proof that dirt existed on the planet way back then. The stuff in your backyard is much fresher. "Most of the dirt you see today is from the past 2 million years," Pavich says. About 2 million years ago, the planet underwent two major changes that drove the formation of new dirt. Global cooling and drying enlarged the deserts, and dust storms redistributed that dirt around the globe. Meanwhile,

glaciers began extending from near the poles, grinding rocks, soil, plants, and anything else into dirt as they moved over the land.

Dirt is still being produced all the time, albeit in much lesser quantities. Beneath the soil's surface, rocks constantly react with rainwater or groundwater and slowly grind together to break down into smaller minerals. So in that respect, dirt really isn't that old. Then again, Pavich notes, a lot of what came out of the Big Bang was essentially dust, which then condensed to form the stars and, later on, planets. "If you think about it," he says, "dirt and its origin are older than the stars."

135

WHAT'S THE BIGGEST THING A CARNIVOROUS PLANT WILL EAT?

FEED ME, SEYMOUR

Carnivorous plants generally stick to a diet of bugs they ensnare. On rare occasions, though, tropical pitcher plants—which drown and break down prey in vase-shaped traps that can be smaller than a little finger or larger than a football—have been found holding the skeletal remains of frogs, geckos, and even small rodents.

That said, chowing down on a vertebrate is incredibly dangerous for the plant, says Barry Rice, author of Growing Carnivorous Plants and the former conservation director for the International Carnivorous Plant Society. It takes a long time for the plant to digest meat, so the hapless critter could rot prematurely, which would kill the trap.

Cutting straight to the heart of what we all want to know, a giant meat-eating plant could still definitely develop a taste for humans. While recovering from a

case of athlete's foot, Rice fed infected skin to Venus flytraps to see if they would eat it. A week later, he was astonished (and a bit appalled) to find barely a trace of his skin remaining in the traps. Healthy skin and internal organs would probably meet the same end, Rice predicts. "I'm still fond of my fingers, though," he says. "So I'm not taking the experiment to the next level."

136

WHY DOES THE CORPSE FLOWER SMELL SO BAD?

ZOMBIE BOUQUET

The better to gross you out, my dear. That distinct scent, somewhere between a rotting pumpkin and a decaying animal carcass, plays an important evolutionary role for the corpse flower, more formally known as *Amorphophallus titanum*, or titan arum. "The plants have learned to mimic those smells for the sake of attracting pollinators," says Ernesto Sandoval, the curator of the University of California at Davis's Botanical Conservatory and proud caretaker of Tammy, Tommy, Ted, and Tabatha, the resident specimens. The titan arum has found its odd evolutionary niche in appealing to flies and carrion beetles. (The plant is one of many species of flowers and fungi that use the smell of death to lure carrion-eating pollinators.) If all goes the corpse flower's way, sometime during the few days that the plant blooms every two years, pollen will rub off on curious insects, which will then fly on to pollinate other titan arum plants in the area.

For help making sure their smell gets around, the flowers crank up the heat. Through a process called mitochondrial uncoupling, the corpse flower (which can grow up to 9 feet [2.7 m] high) takes the energy produced through photosynthesis and, instead of using it to make food, releases the energy as a form of heat. This extra heat—the tip of the plant can become as hot as 100°F (37.8°C)—produces a so-called chimney effect, and the warmth makes the smelly compounds more volatile, allowing them to spread farther in the surrounding environment. The odor can be detected 1 mile (1.6 km) away.

SECRET RECIPE

No one knows precisely which chemicals are responsible for the stink of the corpse flower, but the main odorants are sulfur-containing molecules such as the aptly named putrescine and cadaverine. The combination of these chemicals, along with their concentrations, creates the varied nuances of the scent detected throughout the plant's blooming cycle. Whatever the recipe for the corpse flower's smell, it will gross you out, guaranteed.

137
WHAT HAPPENS TO DISARMED NUCLEAR WARHEADS?

▶ HIGH-STAKES RECYCLING

To conform to the New Strategic Arms Reduction Treaty, the U.S. must reduce its nuclear arsenal to include fewer than 1,550 deployed nuclear strategic warheads. Once the National Nuclear Security Administration's scientists inspect each warhead to determine the best way to take it apart, it goes to Pantex Plant in Texas, the only facility in the U.S. cleared to disassemble nukes.

The conventional explosive used to trigger the nuclear reaction is the most volatile part of the bomb, so it is removed first and burned. Plutonium and highly enriched uranium are removed and shipped to other facilities, where they are later converted into fuel rods for use in nuclear-power plants.

ALL IN A DAY'S WORK

One might suspect that a person who handles nuclear bombs would need a steady supply of antacids. But bombs are recycled in the same slow, reliable, boring sort of way as other technologies—only, while wearing a radiation suit. "That said, you can never lose respect for the material you're handling," says Bron Johnston, the dismantlement manager at the Y-12 facility at the National Security Complex in Oak Ridge, Tennessee, where uranium is handled. "Safety is paramount. It's not like you're working in a bread factory."

CAN WE DISPOSE OF RADIOACTIVE WASTE IN VOLCANOES? 138

▶ SURE, IF YOU LIKE GLOWING RAIN

Dumping all our nuclear waste in a volcano does seem like a nice, tidy solution for destroying the roughly 29,000 tons (26 million kg) of spent uranium fuel rods stockpiled around the world. But there's a critical standard that a volcano would have to meet to properly dispose of the stuff, explains Charlotte Rowe, a volcano geophysicist at Los Alamos National Laboratory. And that standard is heat. The lava would not only have to melt the fuel rods but also strip the uranium of its radioactivity. "Unfortunately," Rowe says, "volcanoes just aren't very hot."

Lava in the hottest volcanoes tops out at around 2,400°F (1,316°C). (These tend to be shield volcanoes, so named for their relatively flat, broad profile. The Hawaiian Islands continue to be formed by this type of volcano.) First of all, that's not hot enough to melt even the zirconium (melting point 3,371°F [1,855°C]) that encases the fuel, let alone the fuel itself: The melting point of uranium oxide, the fuel used at most nuclear power plants, is 5,189°F (2,865°C). It takes temperatures tens of thousands of degrees hotter than that to split uranium's atomic nuclei and alter its radioactivity to make it inert, Rowe says. What you need is a thermonuclear reaction, like an atomic bomb—not a great way to dispose of nuclear waste.

Melting points aside, a volcano probably wouldn't even swallow the material. The liquid lava in a shield volcano pushes upward, so the rods would be unlikely to sink very deep, Rowe says. They wouldn't sink at all in a stratovolcano, the most explosive type, exemplified by Mount St. Helens in the U.S. Instead, the waste would just sit there on top of the volcano's hard lava dome—at least until the pressure from upsurging magma became so great that the dome cracked and the volcano erupted. And that's the real problem.

A regular lava flow is hazardous enough, but the lava pouring out of a volcano used as a nuclear storage facility would be extremely radioactive. Eventually it would harden, turning that mountain's slopes into a barren nuclear wasteland for decades to come. And the danger would extend much farther. "All volcanoes do is spew stuff upward," Rowe says. "During a big eruption, ash and gas can shoot 6 miles [10 km] into the air and afterward circle the globe several times. We'd all be in serious trouble."

139

COULD **TAPPING** THE **PLANET** FOR **GEOTHERMAL ENERGY** COOL DOWN THE **EARTH'S CORE?**

▶ **PLUG IT IN**

Global warming, holes in the ozone layer, and lush golf courses in the desert all reveal mankind's ability to mess with the planet. But the Earth's core, protected by an outer core consisting of some 1,000 miles (1,600 km) of 8,000°F (4,400°C) liquid metal, appears safe from our meddling.

Geothermal energy systems don't drain heat straight from the core. Instead, they pull radiant heat from the crust—the rocky upper 20 miles (32 km) of Earth's surface—either by sucking up pockets of hot water or by circulating water through a closed system embedded in the rock. Power plants use steam from the hot water to spin turbines that make electricity. Geothermal energy generates 7 to 10 billion watts worldwide, which barely accounts for 0.05 percent of global energy consumption and far less than the estimated 44 trillion watts the planet produces.

But drawing energy from the crust won't send it into a deep freeze: Its heat is constantly renewed by the virtually continuous decay of radioactive elements sprinkled throughout it.

WHAT FILLS THE SPACE LEFT IN WELLS WHEN OIL IS EXTRACTED FROM THE GROUND?

140

NOT MORE OIL, UNFORTUNATELY

You might guess that magma or rocks fill the void, but the truth is much more prosaic—water. Petroleum deposits, which are naturally mixed with water and gas, lie deep within the Earth's crust in layers of porous rock, typically sandstone or limestone. (Contrary to what you might imagine, drilling for oil is more like sucking oil from a sponge than from a giant pool of liquid.) At such depths, these liquids are under extremely high pressure. Pump petroleum out, and the pressure in the well drops. Water in the surrounding rock, which is also packed under high pressure, then pushes its way into this low-pressure pocket until the pressure reaches equilibrium. "It's just like digging a hole at the beach, where water in the sand around it flows into the lower-pressure zone of the hole," explains Chris Liner, a professor of petroleum seismology at the University of Houston.

LITTLE EARTHQUAKES

Yes, but not much. Unless you drill in a volcanically active region (which would be unwise for lots of reasons), magma typically flows miles below the deepest oil wells, which tap out around 30,000 feet (9,100 m) down—so there's not much chance of unleashing a lava geyser. And although some shifting of rock and deep sediment can occur, it wouldn't spur a major earthquake. Typical drilling-induced quakes register a magnitude between –2 and –4, which is one-thousandth as forceful as the rumble of a tractor trailer driving by.

142

WILL WIND FARMS MESS UP THE WEATHER BY INTERFERING WITH STORMS AND WIND?

▶ **WATCH OUT, SCARECROW**

Only if you're a cornstalk. Computer models of atmospheric physics suggest that very large wind farms—far larger than exist today—could affect the climate, but only directly underneath the turbines.

The ground temperature below a 10,000-turbine farm might rise by about 1.3°F (0.72°C), according to a simulation described in a 2004 issue of the *Journal of Geophysical Research*. The spinning rotors mix air and pull moisture upward, warming and drying out the air closest to Earth. This could cause grief to the crops planted directly underneath the wind farm (by increasing the need for irrigation), but globally the effect wouldn't add up to much. Simulations led by researchers at the University of Calgary revealed that even if there were enough turbines to produce 17 terawatts of electricity—several times the largest estimate for the next half-century—any local changes would still dissolve into the larger atmosphere.

Today's wind farms are too small (the largest has 627 turbines) to influence the weather at all. And even if they could, energy experts agree that the benefits of wind power would far outweigh the harm. "There is a price to be paid [even for] renewable energy," writes study author Somnath Baidya Roy of the University of Illinois. "[But] when climate change and air quality costs are considered, wind power comes out ahead."

143

CAN I MAKE SNOW WITH JUST MY GARDEN HOSE?

▶ **LOW-BUDGET SNOW DAY**

Sorry, Jack Frost, but setting your puny garden hose to "mist" when it's cold outside isn't going to turn your frozen lawn into a sledding hill. Why not? The problem, in a word, is dirt. Natural snow forms in the upper atmosphere when tiny water droplets adhere to ice crystals or a small speck of dust and then change from supercooled liquid water to solid ice. These crucial dust and ice "nucleation" points are the secret ingredient that's missing from your DIY snow venture, says Matthew Pittman, cofounder of snow-machine manufacturer SNOWatHOME in Connecticut. "Just spraying water in a freezing environment won't make snow," he explains. "The water will not freeze until it makes contact with the ground." And it would be just ice, not nice, fluffy snow.

Professional snow-making machines, which use a proprietary blend of nucleation particles (basically, designer dirt and bacteria), use 150 gallons (567.8 liters) of water a minute to keep the slopes of your favorite ski resort covered with the white stuff. A typical garden hose spews out a paltry 6 gallons (22.7 liters) a minute. So even if you did manage to get your water droplets to form snowflakes before they hit the ground, all you could hope for is a very slow accumulation—and the smallest snowman ever.

WHY DON'T SNOWSTORMS PRODUCE THUNDER AND LIGHTNING?   144

THOR IS HIBERNATING

Nearly every summer rainstorm comes with bursts of thunder and lightning. Yet during even the blusteriest blizzard, there's nary a spark in the air. Thundersnow isn't impossible (though snow-related lightning strikes only six times a year on average in the U.S.), but winter air just doesn't make for prime lightning-forming conditions, says meteorologist Robin Tanamachi of the University of Oklahoma.

During the summer, the lower atmosphere is full of warm, humid air. Above that, it's cold and full of ice crystals. As the warm air rises, carrying water vapor with it, these molecules brush against the ice crystals, and the friction creates an electric field in the cloud—

like when you scuff your feet across a carpet. The ice crystals gain a slight positive charge, and the updraft carries them to the top of the cloud, giving the cloud's bottom a net negative charge. Once the difference between the negatively charged cloud bottom and the positively charged ground becomes great enough, a bolt arcs between them.

But in snowy months, the atmosphere is almost always cold and dry throughout, so there's no updraft that could create friction within the clouds. Wind stirs the molecules and crystals around some, but that action rarely generates the strong electric field it would take to spark lightning.

145

CAN WE PROVE WHETHER GHOSTS AND OTHER PARANORMAL PHENOMENA EXIST?

▶ WHO YA GONNA CALL?

Most researchers still don't consider reports of things that go bump in the night worthy of scientific study, but there are a few who do. For more than a decade, neuropsychologist Jason Braithwaite of the University of Birmingham has been trying to figure out why visitors to Muncaster Castle in England say they see ghosts in rooms and corridors. "People have been having these experiences for hundreds of years, and they deserve a scientific explanation," he says. And although Braithwaite doesn't have an explanation for the supernatural shenanigans, he's found some leads.

His work is based on studies conducted by Michael Persinger of Laurentian University in Canada, who observed several years ago that low-frequency electromagnetic fields (EMFs), typically generated by laboratory magnets, affect the right hemisphere of the brain in a way that makes people believe they're sensing an otherworldly presence. When a specific region of the right hemisphere that controls notions of the "other" is stimulated by EMFs, subjects often report feeling strongly that they are not alone—that there's something, or someone, in the room with them. Using a device he developed with sensors that measure the strength of these low-frequency EMFs, Braithwaite detected unusual electromagnetic activity below the 50-hertz range in and around the castle, a possible explanation for the years of ghost reports.

146 I WANT TO STUDY GHOSTS. WHAT SHOULD I DO?

▶ SEANCE 101

You're in luck! Bothered by the profusion of paranormal investigators who see the scientific method as optional, Jason Braithwaite created the first-ever "ghost school." It teaches anyone who's interested how to investigate paranormal phenomena logically and rigorously. For instance, a thorough investigator might survey the public to zero in on locations where many people report feeling spooked and then investigate those areas to see if magnetic fields or chilly drafts, which can also tweak your senses, might be responsible. "The course isn't about teaching people what to think but how to think," Braithwaite says.

He notes, however, that even with extensive research, it might be impossible to definitively disprove the existence of ghosts. "Factors like magnetic fields may be potential explanations for some sightings," he says. "But others are still mysteries."

WHAT WOULD HAPPEN
IF EVERY ELEMENT ON THE PERIODIC TABLE
CAME INTO CONTACT SIMULTANEOUSLY?

AWESOME MUTANT ELEMENT, HERE WE COME

There are two ways you could theoretically go about testing this: Try and combine single atoms of each element, or bring together fistfuls of each element in its natural state and see what happens. Neither option is practical. One requires the energy of dozens of Large Hadron Colliders. Another could yield a cauldron full of flaming plutonium. Both, however, would probably create carbon monoxide and a pile of rust and salts rather than a cool Frankenstein element.

FIRST, THROW SOME ATOMS TOGETHER, RIGHT?

If you toss single atoms of each element into a box, they won't form a supermolecule containing one of everything, explains Mark Tuckerman, a theoretical chemist at New York University. Atoms consist of a nucleus of neutrons and protons with a set number of electrons zooming around them. Molecules form when atoms' electron orbitals overlap and effectively hold the atoms together. What you get when the atoms are left to their own devices, Tuckerman says, will be influenced by what's close to what. Oxygen, for example, is very reactive, and if it is closest to hydrogen, it will make hydroxide. If it is nearest to carbon, it will make carbon monoxide. "That random reactive nature applies to pretty much all elements," Tuckerman says. "You could run this experiment 100 times and get 100 different combinations." Certain elements, such as the noble gases, wouldn't react with anything, so you'd be left with those and a few commonly found two- and three-atom molecules.

OK, WHAT IF I SMASHED THEM TOGETHER SUPER HARD?

You could try to get all the atoms to hit each other at exactly the same time by running them through a particle accelerator. Still, ramming the atoms together at 99.999 percent the speed of light—the top speed of particles in the Large Hadron Collider—might fuse a few nuclei, but it won't make that cool Frankenstein element, either. More likely, they would meld into a quark-gluon plasma, the theoretical matter that existed right after the universe formed. "But that would last for a fraction of a second before degrading," Tuckerman says. "Plus, you'd need 118 LHCs—one to accelerate each element—to get it done."

FINE, WHAT IF I JUST THREW A HANDFUL OF EACH ELEMENT TOGETHER?

The other approach, as explained by John Stanton, the director of the Institute for Theoretical Chemistry at the University of Texas, would be to toss a pulverized chunk of each element or a puff of each gas into a sealed container and see what happens. No one has ever tried this experiment, either, but here's how Stanton thinks things would play out: "The oxygen gas would react with lithium or sodium and ignite, raising the temperature in the container to the point that all hell would break loose. Powdered graphite carbon would ignite, too. There are roughly 25 radioactive elements, and they would make your flaming stew a little dangerous. Flaming plutonium is a very bad thing. Inhaling airborne radioactive material can cause rapid death." But once things calmed down, Stanton says, the result would be just as boring as the atoms-only scenario. Carbon and oxygen would yield carbon monoxide and carbon dioxide. Nitrogen gas is very stable and would remain as is. The noble gases wouldn't react, nor would a few of the metals, like gold and platinum, which are mostly found in their pure forms. The things that do react would form rust and salts. "Thermodynamics wins again," he says. "Things will always achieve equilibrium, and in this case, that's a mix of common, stable compounds."

148

I'VE HEARD THAT THE EARTH'S ROTATION IS SLOWING. HOW LONG UNTIL DAYS LAST 25 HOURS?

DAYLIGHT SAVINGS

We could all use an extra hour in the day, but clocks won't need to be redesigned anytime soon. The time the Earth takes to make a complete rotation on its axis varies by about a millionth of a second per day, says physicist Tom O'Brian of the National Institute of Standards and Technology. While some days are shorter than average, the planet's rotation shows a long-term slowing trend, which will, indeed, ultimately lead to a longer day.

WINDING DOWN

Scientists have reliable data on the Earth's rotational speed, based on recorded observations of the sun's position in the sky during solar eclipses going back some 2,500 years. Although the planet's rotational rate hasn't declined smoothly over that period, the average day has grown longer by between 15-millionths and 25-millionths of a second every year. Even at the faster rate, it would take at least 140 million years before the Earth's rotation slowed enough to necessitate a 25-hour day.

You don't need to worry about having to add another day to your calendar, either. Although the planet's rotation around its own axis is lagging slightly, we're orbiting the sun as quickly as ever, and that shows no signs of slowing down.

▶ **IT'S GETTING STUFFY IN HERE**

Here's why you might be worried: Burning oil, coal, gas, wood, and other organic materials requires molecular oxygen, the O_2 we breathe, to break carbon-hydrogen bonds and release energy. This reaction, better known as combustion, also pairs each broken-off, positively charged carbon atom with two negatively charged oxygen atoms, forming carbon dioxide, or CO_2.

Although that does cut into the amount of O_2 in the atmosphere, there's no need to stockpile oxygen tanks in your basement just yet. Nitrogen accounts for 78 percent of the gas in the atmosphere, but molecular oxygen is the runner-up, at 20.94 percent. The leftover 1 percent and change falls into the "other" category, predominantly water vapor but also argon and hydrogen gas; CO_2 accounts for just 0.04 percent of what we breathe.

Because of this relative bounty of atmospheric oxygen, scientists such as Pieter Tans of the National Oceanic and Atmospheric Administration don't fear that carbon emissions are eventually going to cut off our oxygen supply. "Even if we were to burn another 1,000 billion tons (900 trillion kg) of fossil fuels, we would only decrease the oxygen in our atmosphere to 20.88 percent," he says. For comparison, we burn just 7 billion tons (6.35 trillion kg) each year worldwide. And even then, the other effects that action would have on the environment—more particulate pollution, hotter temperatures—would be much, much worse than oxygen depletion.

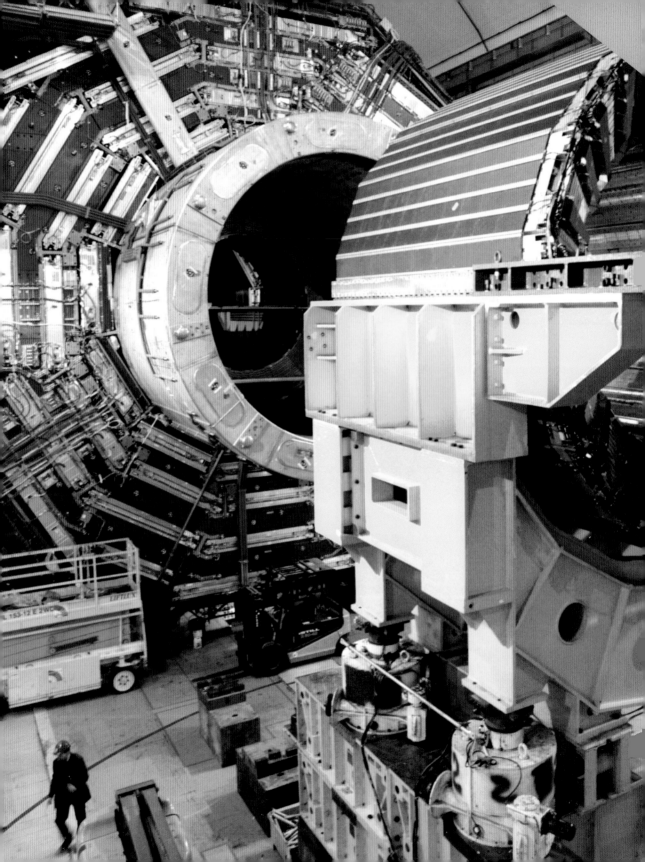

COULD THE LARGE HADRON COLLIDER CAUSE POWER OUTAGES IN EUROPE?

IT'S LIKE A GIANT HAIR DRYER

Everyone's favorite accelerator is definitely an energy hog. Inside a ring stretching 17 miles (27 km) buried near Geneva, Switzerland, the Large Hadron Collider is pushing the boundaries of science and increasing our knowledge of physics as it speeds two beams of protons in opposite directions to energies of seven tera-electron-volts. That takes a lot of juice.

At its peak, CERN, the European physics lab that houses the LHC, consumes 180 megawatts of power, with the accelerator burning through 120 of that total. The biggest electricity suck is the cryogenic system, used to chill 7,000 superconducting magnets to a temperature just above absolute zero. This system draws 27.5 megawatts to steer the proton beams along a circular path. Then come the four major detectors—the machines that actually read the collisions between protons—which draw a combined 22 megawatts. Factor in the powerful electric fields required to accelerate the tiny particles up to nearly the speed of light, and it's not all that surprising that when the LHC is running, CERN drinks up as much energy as a small city.

Although such energy consumption might seem like an undue burden on the local power grid, LHC machine coordinator Mike Lamont says it's actually a 9 percent decrease compared with CERN's previous accelerator, the Large Electron-Positron Collider. And CERN's total power consumption is just 10 percent of what Geneva usually uses, so the risk of cutting off power to the locals is fairly slim.

THERE GOES SUMMER BREAK

The real concern for CERN is the electric bill. The $4-billion price tag for the machine is merely the up-front cost—running the accelerator is expensive, too. The potentially astronomical electricity bill forces the lab to shut down the LHC in the winter, when rates are highest. Working in the summer may not be a very European way to conduct one's affairs, but the secrets of the universe are surely more important than a leisurely August.

151

IS THE **ROCK-CONCERT** LIGHTER SALUTE BAD FOR THE ENVIRONMENT?

▶ **JUST HOLD UP YOUR CELL PHONE INSTEAD**

First, for the uninitiated, an explanation of the lighter salute: You're at a concert. The music slows, the first chords of a power ballad begin, and hundreds of disposable lighters illuminate the audience like so many sequins on a vest. Three or four (or ten, if it's a particularly long solo) minutes later, the song ends, you pocket the Bic, and get back to headbanging.

But fear not, "Free Bird" devotees. Lighting up en masse isn't all that bad for the environment. The butane in disposable lighters is a compound made up of carbon and hydrogen; as it burns, these elements combine with oxygen to produce carbon dioxide and water vapor. A typical lighter releases about 237 milligrams of carbon dioxide per minute. If 1,000 Night Ranger fans burned their lighters during the entirety of "Sister Christian," they would collectively release about 2.6 pounds (1.2 kg) of CO_2. Compare that with the 26,900 pounds (12,200 kg) a typical power plant produces in a minute. In fact, if you were to give a one-minute salute with each of the 1.46 billion lighters that Bic sells annually, the amount of CO_2 you'd create would equal only 28 minutes of said power plant's emissions. So salute those ballads fearlessly, and rock on.

REDUCE YOUR NICOTINE FOOTPRINT

You may find this hard to believe if you're standing near a swarm of chain smokers, but most scientists think the trace amounts of carbon dioxide and other pollutants in cigarette smoke have, at most, a negligible effect on the climate. "In fact," theorizes John M. Wallace, a professor at the University of Washington's climate-research department, "it might even counteract global warming by an equally minuscule amount, because the white particulate matter in smoke would reflect some of the sun's energy, thereby minimizing heat."

SMOKING KILLS WHALES

Not so fast. The thing is, the smoky end-product is not the entire story. Tobacco must be grown, and that process puts a serious hit on the environment. The plant itself is very demanding, absorbing six times as much potassium from the soil as most crops do. In some developing nations, farmers grow tobacco until the soil is useless and then clear-cut forests for fresh land. In those areas, 600 million trees are felled and burned annually to dry and cure tobacco leaves. Additionally, 4 miles (6.4 km) of cigarette-width paper an hour is used to wrap and package cigarettes.

Setting aside the pollution generated from the cigarette manufacturing process, just the loss of this many carbon dioxide–absorbing trees leaves at least an extra 22 million net tons (20 billion kg) of CO_2 in the atmosphere. This is the rough equivalent of burning 2.8 billion gallons (10.6 billion liters) of gasoline.

The damage isn't confined to the air, either. According to common estimates, tobacco companies produce a yearly total of 5.5 trillion cigarettes—approximately 900 for each person in the world. Of those, 4.5 trillion have nonbiodegradable filters that are tossed away, representing as many as one out of every five pieces of litter. Cigarette butts require months or even years to break down, releasing almost 600 chemicals into the soil.

So although most scientists believe that the act of smoking itself has a zero net effect on global warming on its own, secondhand smoke appears to be a minor annoyance compared with the larger damage cigarettes do to the planet.

HOW MIGHT THE WORLD END FOR US?

ASTEROID STRIKE

Texas-sized asteroids make for fun summer blockbusters, but when it comes to long-term damage, they're not the most menacing threat out there. Still, that's not saying being hit by a hurtling ball of rock would be very much fun. Sixty-five million years ago, an asteroid about 6 miles (10 km) wide slammed into Mexico's Yucatan peninsula, likely causing a mega-tsunami, setting off earthquakes and volcanic eruptions, and blanketing the atmosphere with dust particles for years. The Chicxulub impact, as it's known today, is widely thought to have led to the extinction of the dinosaurs. So, yeah, keep an eye out for the asteroids.

VOLCANIC UPROAR

Underwater volcanoes could spew hydrogen sulfide and carbon dioxide into the atmosphere, which many scientists think caused the worst mass extinction in history 250 million years ago, which killed 90 percent of all marine species and 70 percent of land animals. But there's an upside even to this: Bacteria and other microorganisms thrive in these conditions, so there's a good chance of life hitting the restart button. Just, you know, without us.

DRASTIC CLIMATE CHANGE

A quick, disastrous change in Earth's climate could easily end our terrestrial party early. The UN's Office for the Coordination of Humanitarian Affairs estimates that, already, up to 70 percent of disasters worldwide are climate related, affecting about 2.4 billion people over the last decade. If, as most scientists believe, global warming is set to escalate over the next century, drastic weather events like floods, droughts, and hurricanes are likely to become ever more commonplace. In the worst-case scenarios, this could lead to massive crop failures, shortages of drinking water, and widespread extinctions—including, perhaps, our own.

IS THAT ALL I HAVE TO WORRY ABOUT?

Nope! Even leaving out such sci-fi perennials as alien invasion, total nuclear war, or the zombie apocalypse, the universe is full of other fascinating surprises. For instance, lurking at the edge of our galaxy are giant molecular dust clouds—agglomerations of hydrogen gas, small organic molecules, and minerals—roughly 150 light-years across. If our solar system hit one, it would take 100,000 years for us to pop out on the other side.

During that time, the dust would accumulate in our atmosphere and block out nearly all light from the sun. The oceans would freeze over and terrestrial plants would die off, leading to a near-total extinction of life, says Alexander Pavlov, a NASA-affiliated astrobiologist at the University of Arizona. The good news? We probably won't hit one for at least 40 million years. So there's lots of time for you to get your emergency kit together.

155

IS THERE REALLY **NO WAY** TO **KEEP** GEESE OUT OF A JET ENGINE?

▶ **THIS GOOSE IS COOKED**

Unfortunately, there's pretty much no way to protect jet engines from geese or other large birds. In fact, fastening some sort of shield over a jet engine could actually make things worse.

From 1997 to 2007, reported animal strikes per commercial flight in the U.S. more than doubled, reaching 7,600 (mainly glancing shots) over 54 million flights in 2007, according to Richard Dolbeer, former chairman of the volunteer organization Bird Strike Committee USA. Since most bird strikes occur less than 500 feet (150 m) off the ground, experts blame the growing numbers of Canada Geese and other birds near airports. When a bird is sucked into an engine, the carnage can jam up the turbine and stop the engine. Famously, a flock of geese shut down both engines of a 150-passenger Airbus A320 right after takeoff from New York's LaGuardia airport in January 2009, forcing the pilot to make an emergency landing in the Hudson River.

Although one might assume that airlines could shield jet engines with a screen or a set of bars, it's simply not practical, says Russell DeFusco, vice president of Bird Aircraft Strike Hazard and Wildlife Management Consultants (BASH, Inc.). An average goose weighing 12 pounds (5.4 kg) hits a plane traveling 150 miles per hour (241.4 kph) on takeoff. That's roughly the same force as dropping a grand piano from the second story of a building, says Matthew Perra, a spokesman for engine manufacturer Pratt & Whitney. Beyond the fact that any screen would create turbulence and inhibit free-flowing air from entering the engine (thus weakening its thrust), if the screen broke on impact, it too could be sucked into the engine and cause even more damage, DeFusco says.

Per Federal Aviation Administration (FAA) regulations, before a new engine model can be approved, it must prove that it can safely shut down after ingesting a bird carcass weighing 4 pounds (1.8 kg). This test might not be rigorous enough, however—the largest Canada Geese tip the scales at 14 pounds (6.4 kg).

Bird-repelling techniques now in use at airports include draining fowl-friendly ponds and scaring off birds with firecrackers. The FAA is also working on bird-detecting radar, much like the one Kennedy Space Center has used since a space shuttle's 2005 run-in with a vulture.

WOULD MY DELOREAN FLY IF I POPPED ITS GULL-WING DOORS AND FLOORED IT?

156

▶ BACK TO THE FUTURE

The best bet for getting a DeLorean airborne would be to get a hover conversion in 2015, but we turned to Diandra Leslie-Pelecky, author of *The Physics of NASCAR*, to see if it's possible to fly it the old-fashioned way.

For any vehicle—airplane or car—to fly, there needs to be some force pushing it up so that it can overcome gravity. Airplane wings are specifically designed to create just such a force. As a plane moves forward, the wings push air down, and because for every action there's an equal and opposite reaction, this action creates an upward force on the wing, called lift.

Cars, on the other hand, are designed to fight lift, and your DeLorean's "wings" won't begin to make up the difference. For one, the doors aren't shaped or angled in a way that would push enough air downward to create a significant upward force. They're relatively short, so there isn't much surface to push up on anyway, and they don't protrude straight out, which is ideal for flying. Worse still, the doors' bulky interior padding and the side-view mirror would create a good deal of drag. Not to mention that the doors would probably break off if forced to support the entire weight of the car.

Another knock against a flying DeLorean is that it's very boxy, so even with the doors shut, its 20 square feet (1.8 m²) or so of frontal area (the part of the car that hits the most air) is greater than on most cars its size. Plus, when you open the gull-wing doors, you're inviting air into the car, which creates a lot of drag resistance. A DeLorean's top speed is 105 miles per hour (169 kph) with the doors closed, which isn't nearly enough to overcome this. Even if you installed a race car's engine and could get the car going 200 miles per hour (322 kph), it still wouldn't take off.

But that's probably a good thing. If you did somehow manage to get the car airborne, you'd have another problem: Once the tires left the road, there would be no force pushing the car forward. Drag—and gravity—would take over, you'd slow to the point where lift couldn't keep you aloft, and not even a flux capacitor could prevent you from crashing back to the ground.

157
HOW MANY WAYS CAN A CHESS GAME UNFOLD?

▶ **WE'RE STICKING WITH CHECKERS**

Almost nothing looks more orderly than chess pieces before a match starts. The first move, however, begins a spiral into chaos. After both players make a move, 400 possible board setups exist. After the second pair of turns, there are 197,742 possible games, and after three moves, 121 million. At every turn, players chart a progressively more distinctive path, and each game evolves into one that has probably never been played before.

According to Jonathan Schaeffer, a computer scientist at the University of Alberta who demonstrates artificial intelligence using games, "The possible number of chess games is so huge that no one will invest the effort to calculate the exact number." Some have estimated it at around $10^{100,000}$. Out of those, 10^{120} games are "typical": about 40 moves long with an average of 30 choices per move.

To put all this in perspective, there are only 10^{15} total hairs on all the human heads in the world, 10^{23} grains of sand on Earth, and about 10^{81} atoms in the universe. The number of typical chess games is many times as great as all those numbers multiplied together—an impressive feat for 32 wooden pieces lined up on a board.

158

WHAT'S THE FEWEST MOVES IT TAKES TO SOLVE A RUBIK'S CUBE?

▶ IT'S NOT JUST A FANCY PAPERWEIGHT

Most people need hours to bring every side of a Rubik's Cube into monochromatic perfection, but an omniscient player would be able to solve any position of the cube in 20 moves or less. A team of mathematicians, engineers, and other researchers created a computer algorithm that successfully solved all 43,252,003,274,489,856,000 possible positions of the cube, finally proving that no position, however impossible-looking, would take more than 20 moves to solve.

The team was able to determine this number (which they nicknamed "God's Number") by dividing the possible positions into sets and reducing the number of sets through symmetry (you can avoid solving some positions because they are mirror-images of another position). Through this method, they reduced the number of sets that actually had to be solved to a more manageable 55,882,296. The program the team wrote to solve the positions took around 20 seconds per set, and using about 35 CPU-years of computer time donated by Google, the team cracked the Rubik's Cube within a few weeks. Next time you're stuck, just ask your computer for help. Sure, it's embarrassing, but it beats peeling off and rearranging the stickers.

CAN MICROWAVE TECHNOLOGY BE USED TO MAKE FOOD COLD?

INSTANT POPSICLES

Microwaves can transform a frozen pizza into hot, melted goodness in four minutes flat, but they can't rescue your ice-cream sundae. This is because, in order to cook food, a microwave oven converts voltage into high-frequency electromagnetic microwaves. The molecules in food—especially water and fat molecules—absorb this energy and wiggle at high speeds, causing them to heat rapidly and warm the surrounding food. Although quickly turning leftovers cold would be handy, this is a one-way operation, explains David Pozar, a microwave expert and professor emeritus at the University of Massachusetts. Microwaves can only speed atoms up, not slow them down.

ICE-COLD LASERS

Scientists do have a high-tech method for slowing atoms, however: lasers. Shoot a moving atom with a laser, and it will absorb the laser's photons and re-emit them every which way, causing the atom to hold nearly still. Placing an atom at the junction of multiple beams can slow its momentum in all directions, decreasing its energy and cooling it.

This drops an atom's temperature a couple hundred degrees—much colder than anything you'd want to put in your mouth—in less than a second. But because it works most efficiently on low-density gases of atoms of a single element, physicist Mark Raizen of the University of Texas doesn't think it will be useful for cooling food anytime soon. "Not unless you can subsist on a thousand sodium atoms," he says.

▶ POWER LUNCH

If you turn your computer off for your lunch hour you will, in fact, save some energy. But those modest energy gains might come at the expense of your computer's longevity. To figure out just how much energy an average computer consumes during its various states of use, we asked Harvard University physicist Wolfgang Rueckner to run a few tests on his 2005 iMac G5. While starting up and shutting down, the machine gobbled about 130 watts (a measure of the amount of electricity used at any instant). It consumed 92 watts sitting idle and whispered along at an efficient 4 watts in sleep mode. Turned off, it sipped 2.8 watts because it was still plugged into an outlet. Adding in the consumption spikes that occur while shutting down and starting back up, the electricity the computer uses while turned off for an hour is only very slightly less than what it consumes while sleeping.

Given these numbers (assume it's even higher for a computer running Windows, which requires more processing power than a Mac operating system), if 20 people in your office turned off their computers for lunch, you'd collectively save 24 watts during the hour—about what it takes to light a standard compact fluorescent bulb. At the U.S. Department of Energy's 2010 average commercial-energy price of 10.4 cents per kilowatt-hour, you'd collectively save—drum roll, please—one quarter of a cent a day.

OFF IS OUT

But you'd lose that scant savings over time, Rueckner says, because your hard drive would wear out more quickly from all the spinning it would do while booting up or down. It would take a lot more quarter-cents to justify the cost of a replacement drive.

Increasingly efficient computers with improved power-management settings will narrow the off/sleep gap. Bruce Nordman, a researcher at Lawrence Berkeley National Laboratory, says that even though turning a computer off will never actually waste power, he notes that "off" is a very twentieth-century idea.

161

WHY DOES **ORGANIC MILK** HAVE A **LONGER SHELF LIFE** THAN THE **REGULAR KIND?**

▶ **IT'S JUST MORE VIRTUOUS**

It all has to do with where the cow was milked. "Organic milk often has to travel thousands of miles to reach distribution points," says Dean Sommer, a cheese and food technologist at the Wisconsin Center for Dairy Research at the University of Wisconsin. To survive the journey and leave time to spare in the fridge, farmers usually pasteurize organic milk at higher temperatures than conventional milk.

Nearly all milk is pasteurized, or heat-treated, to kill off disease-causing microbes. Heating organic milk upward of 200°F (93.3°C) instead of at the typical 161°F (71.7°C) destroys more of the organisms responsible for spoiling milk. With those bugs knocked out, organic milk lasts 25 to 40 days longer than the ordinary stuff.

But there's a catch. The extra heating is expensive and can give the milk "a cooked or scorched flavor," Sommer says. And be sure to drink up once you crack the carton. Exposed to air, organic milk goes bad just as quickly as any other milk.

WHY DOES COKE FROM A GLASS BOTTLE, A PLASTIC BOTTLE, AND AN ALUMINUM CAN TASTE DIFFERENT? 162

ALL COKES ARE CREATED EQUAL

It doesn't. That's what Coca-Cola's spokespeople say, anyway. "The great taste of Coca-Cola is the same regardless of the package it comes in," they insist. Rather, they say, "The particular way that people choose to enjoy their Coke can affect their perception of taste." Sure, most people would agree that the cola is indeed delicious and refreshing in any vessel, and pouring it into a glass or serving it over ice could influence the way you experience its flavor. But is it possible that the subtle variation in taste that some notice among aluminum cans, plastic bottles, and glass bottles is more than just a psychological effect of their soda-consumption rituals?

Given that the formula is always the same, yes, according to Sara Risch, a food chemist and member of the Institute of Food Technologists. "While packaging and food companies work to prevent any interactions, they can occur," she says. For example, the polymer that lines aluminum cans might absorb small amounts of soluble flavor from the soda. Conversely, acetaldehyde in plastic bottles might migrate into the soda. The FDA regulates this kind of potential chemical contact, but even minute, allowable amounts could alter flavor.

163 HOW DO I GET THE TASTIEST SODA?

JUDGE A COKE BY ITS BOTTLE

Your best bet for getting Coke's pure, unaltered taste is to drink it from a glass bottle, the most inert material it's served in. Even that's not a sure bet, though. Coca-Cola maintains strict uniformity in processes in all of its worldwide bottling facilities, but it concedes that exposure to light and the length of time the product sits on store shelves may affect the taste. So yeah, the packaging might mess with Coke's flavor, but we'll still take it any day over New Coke.

164

HOW DO I TAKE A BLINK-FREE GROUP PHOTO?

▶ **SAY CHEESE . . . AGAIN**

Sitting for those class pictures in elementary school was always a chore, especially considering that half the kids ended up looking asleep, their faces immortalized mid-blink. It turns out that the seventh grade's photographic legacy might have been in better hands with a physicist behind the lens rather than Bob from Sears.

Frustrated with an excess of closed eyes in her photos, Nic Svenson, a communications officer at CSIRO, Australia's national science agency, enlisted the aid of physicist Piers Barnes to develop a mathematical formula for calculating the number of photographs one ought to take to produce a group shot sans blinkers. Barnes's rule—$1/(1 - xt)n$—accounts for a person's average number of blinks per second (x), the camera's shutter speed plus the duration of an average blink (t), and the number of people in the group (n). Simply plug in your numbers and snap away.

Because most photographers can't carry out tricky algebra in their head, Barnes adapted the formula to an easier rule for figuring out the number of photos to take of groups smaller than 20: In good light, divide the number of people by three; in darker conditions, divide by two (with the shutter open longer for a better exposure, there's more time for blinks to creep in).

For their efforts, Barnes and Svenson won the 2006 Ig Nobel prize—an annual award that recognizes zany scientific research—for mathematics. Although this bit of math didn't earn a real Nobel, it's an insight you can actually use. No word yet on how to calculate the number of photos needed before everyone in the group looks happy, but we're guessing they don't make memory cards large enough.

WHY DO STRINGS OF HOLIDAY LIGHTS END UP IN GIANT KNOTS?

▶ **SANTA'S CURSE**

A century of research on knot theory has confirmed what every neighborhood house-decorating champion already knows through cruel experience—that knots are an unstoppable force of nature. The ingredients of a knot are simple: a string with one loose end, one loop, and some movement to push the end through the loop.

Most people coil their lights for storage, which creates several loops for loose ends to dart through as the lights get moved around the garage. Andrew Belmonte, a mathematics professor at Pennsylvania State University, jokingly suggests that one way to avoid the motion that creates knots is to store your lights by hanging them from rafters "like sausages."

Under certain conditions, however, there's no stopping the knots. University of California at San Diego biophysicist Douglas Smith and research associate Dorian Raymer tumbled strings of various lengths one at a time in a box, like a sock in a clothes dryer. Within seconds, each string tied itself into a knot. After 3,000 tosses, the researchers identified 120 types of knots, and their computer simulations predicted that if the experiment continued indefinitely, they would create an infinite number of supercomplex knots. The lesson learned: shaking your tangled mess in frustration will only make a bigger, bulkier knot.

Yet the experiment also revealed a trick for preventing knots from forming. When the strings were packed snugly in the box, the knots were much less severe. "There has to be a little motion to make knots," Smith says. "It's not magic."

So if you don't have rafters for the sausage-hanging option, follow these simple storage steps: Eliminate loose ends by plugging the two ends of each strand into each other, box them in a tight squeeze, and put them in a spot in your garage where they won't get jostled. You might be pleasantly tangle-free next holiday season.

166
WHAT'S THE DIFFERENCE BETWEEN
ARTIFICIAL AND NATURAL FLAVORS?

▶ **JUST LIKE MOM USED TO MAKE**

Deciding whether to go with barbecue-flavor potato chips or salt-and-vinegar can be tough enough without having to choose between brands made with "natural flavors" and ones that are "artificially flavored." Natural flavors, you might think, are derived from the pure essence of a food's flavor and thus are more authentic. But the term "natural" is misleading.

The U.S. Food and Drug Administration requires that natural flavors come from a natural material, but that's a broad distinction. It usually means chemically treating natural molecules, or making flavors from plant or bacteria by-products, which chemists then tinker with. The strawberry taste of your naturally flavored ice cream? That probably started as a bacterial protein.

Artificial flavors, on the other hand, are just what you'd expect: taste bud–stimulating chemicals concocted from scratch in labs. Although natural flavors have the potential to be more accurate and contain layers of flavor, mucking with bacteria is expensive and the results are inconsistent. Controlling every step of a flavor's development, as chemists do with artificial flavors, costs less and often hits closer to the mark.

Flavor chemists can further enhance artificial flavors by working to stimulate your nose as well as your tongue. "Aroma is often the dominant factor in flavor perception," says Anuradha Prakash, a professor of food science at Chapman University and a spokesperson for the Institute of Food Technologists. "Flavorists can mix compounds with similar tastes but different aromas to maximize artificial flavor."

Despite the healthy sound of the phrase, natural flavors aren't any better for you than man-made ones. In fact, in most cases your body can't even differentiate between the two. Still deciding on those barbecue chips? You'll save a few cents with the artificially flavored variety, and they might even taste more like the real thing.

▶ DOUBLE BACON CHEESEBURGER, PLEASE

Our love affair with a big, juicy sirloin steak goes back more than two million years. Then, even a full day of foraging in the relatively vegetation-bare African savannah yielded too few nutrient-rich plants. So humans began hunting protein-packed animals, explains Katharine Milton, a physical anthropologist at the University of California at Berkeley. "You can't walk dozens of miles every day with your gut full of straw," she points out.

In addition to being high in protein, which helps build muscles, meat is a good source of sodium—an important ion for cell communication and the transmission of signals through the nervous system. It's also a plentiful source of fat, which in addition to giving meat its juicy texture is critical for the absorption of certain vitamins, supplies fatty acids essential for making hormones, and provides cushioning for delicate organs.

A well-cooked steak tastes so delicious to our modern selves because, over time, our bodies learned to associate the taste of meat's fat, protein, and salt with its implicit nutritional value, says Craig Stanford, a biological anthropologist at the University of Southern California.

Dietary requirements aside, there are taste factors on the molecular level that make a heaping mound of barbecued ribs irresistible. Meat is loaded with molecules called ribonucleotides and the amino acid glutamate (which is also abundant in other food, such as parmesan cheese and tomatoes.) When these two react, the product stimulates receptors on your tongue and creates a strong taste known as umami, generally described as "meaty" or "savory," which was discovered by Japanese scientists a century ago. That might explain why a few pieces of bacon make such an excellent addition to a lettuce-and-tomato sandwich.

WHAT ARE SOME EXCITING WAYS
TO LIVE DANGEROUSLY?

MAKE BLACK POWDER

To concoct this primitive form of gunpowder, just mix simple chemicals, like charcoal and saltpeter. Oxygen molecules found in saltpeter will react with the sulfur in charcoal to unleash a torrent of energy, smoke, and noise. Try the recipe in William Gurstelle's great book on dangerous living, *Absinthe & Flamethrowers: Projects and Ruminations on the Art of Living Dangerously*. (Gurstelle supplied all these ideas, by the way.)

BLACK POWDER RECIPE

Ingredients:
- Several small chunks of wood
- Heavy-duty aluminum foil
- A medium-sized nail
- Regular charcoal briquettes
- A grill
- Charcoal-lighting method
- Long-handled lighter
- Grill tongs
- Mortar and pestle

CONSTRUCT A FLAMETHROWER

A propane accumulator flame cannon is a relatively easy project—just make sure your pipe joints are tight, limit fuel pressure, and operate it at a steady level. Still, you'll want to keep a fire extinguisher handy. After all, it's hard to take on future challenges without all your fingers and toes.

BUILD A ROCKET

Nothing impresses a cute geek girl like building her a high-performance rocket out of stuff from your garage. All it takes is the proper mixture of fuel (granulated sugar) and oxidizer (tree-stump remover).

EAT FUGU

Every adventuresome gastronome must someday dine on fugu sushi. But you should always have an expert serve the delicate tiger puffer-fish flesh, since some parts contain tetrodotoxin, a deadly neurotoxin that has no known antidote.

CRACK A WHIP

A whip transfers momentum from your arm to its tip, creating a thrilling mini sonic boom at 770 mph (1,240 kph). Eye patches and novice whip-crackers go hand in hand, so be sure to wear goggles.

169

SHOULD I BE WORRIED ABOUT
ELECTROMAGNETIC PULSES
DESTROYING MY ELECTRONICS?

ULTIMATE SUNBURN

It depends on the source of the pulse. Electromagnetic pulses (EMPs) large enough to cause you trouble come in two varieties: those produced by the sun and those created by a nuclear bomb or another military-grade device. With the sun-related variety, specifically coronal mass ejections (CMEs), your gear will probably be fine. But a really large CME could take down the power grid, says Bill Murtagh, the program coordinator of the National Oceanic and Atmospheric Administration's Space Weather Prediction Center. Power lines transmit electricity as an alternating current, but a pulse from a CME could potentially introduce a direct current into the system, says Luke van der Zel, a technical executive at the nonprofit Electric Power Research Institute. This could cause transformers to overheat and work sluggishly or fail altogether.

Despite the grid's numerous built-in safeguards, if enough transformers go down, they could take large chunks of the grid with them. The only way to get it running again would be to replace all the damaged gear. On the bright side, although CMEs have been known to put satellites out of commission, our atmosphere deflects most of the energy, so the radiation is too diffuse by the time it reaches your electronics to destroy them.

A man-made EMP poses a greater threat. If one goes off in your neighborhood, there's a significant risk that the concentrated pulse will induce extra voltage in the circuit-board components, frying them for good.

170

IS THERE ANY WAY
TO PROTECT MY GEAR?

RENEW YOUR WARRANTY

The best bet for protecting your electronics is to store them in a Faraday cage: a cube of interweaving metals, preferably copper and steel a quarter inch (6 mm) thick, which together can act as an electromagnetic shield. Like in a lightning rod, the copper attracts electricity while the steel absorbs magnetic pulses. A cage that's large enough to hold all your favorite gadgets—your cell phone, TV, computer, and so on—runs in the neighborhood of $15,000. An EMP could also crash the power grid, so you might want to spring for an extra cage to protect your generator, too. Of course, if a nuclear bomb goes off in your neighborhood, you'd probably have bigger worries than whether you've backed up your laptop.

▶ **NO TEXTING IN BED**

Cell phones have been getting bad press for years. First there were the brain-cancer rumors. Then came evidence linking usage with higher stress levels and poor driving. But perhaps the scariest thing for guys is the claim that the devices are doing serious damage to male fertility. Scientists at the Center for Reproductive Medicine at the Cleveland Clinic sampled 361 men and found that the more time the men spent on their cell phones each day, the lower their sperm count.

On the face of it, the results are alarming for any cell phone user. Men who used their cell phones for more than four hours daily had a sperm count 41 percent lower than those who never used them, according to the researchers. Even guys using their phones for just two hours a day saw their troops reduced by 20 percent.

The mechanism by which the damage occurs is fuzzy, says study leader Ashok Agarwal. One possibility is that the electromagnetic radiation that cell phones emit—during a call as well as at rest—might affect the region of the brain responsible for initiating testosterone production. A greater likelihood is that radiation-producing phones—often clipped to a belt or kept in a pants pocket by men who use their phones heavily—affect developing sperm; independent studies have shown that radio waves with frequencies similar to those emitted by cell phones can damage sperm's DNA.

But don't put on the lead underpants just yet. The study didn't control for age, stress, or other contributing factors, and most fertility doctors wouldn't consider the sperm-count levels of even the most prolific gabbers to be clinically abnormal. "I haven't stopped using my phone," says Andrew La Barbera, scientific director of the American Society for Reproductive Medicine.

"To prove the cause and effect is going to be hard," admits Agarwal, who is conducting a more stringent follow-up study. "Perhaps those who are using a cell phone many hours a day should keep it away from their body."

172
WHICH OFFERS BETTER AUDIO QUALITY,
VINYL OR DIGITAL RECORDINGS?

▶ BROKEN RECORD

Sorry, vinyl aficionados, but CDs most accurately capture the clarity of musical performances. If you look at the grooves of a standard long-play record, or LP, through a microscope, you'll see that each is filled with what look like rolling hills. These are, in fact, an extremely close replication of the shape of the sound waves from the musician's instrument. But because the needle that carves the groove is shaped slightly differently from the needle that reads it, the LP will never sound exactly like the original performance. (Not to mention that changes in temperature and humidity warp vinyl over time.)

The mathematical data encoded on a CD, however, is a nearly exact representation of the original sound. Comparing an LP and a CD made from the same microphone signal, the LP's groove must perfectly match the signal to sound close to CD-quality, which is almost impossible, says Stanley Lipshitz, who studies electro-acoustics and digital-signal processing in the Audio Research Group at the University of Waterloo in Canada.

Even so, some audiophiles claim to hear a natural sound, vaguely described as "musical warmth," when listening to vinyl. What they're hearing, Lipshitz says, is most likely the deficiencies of the record player. Sound waves from the speakers and the needle's rise-and-fall passage over the grooves cause the LP to vibrate. The needle picks up these extra vibrations and adds them to the music, creating the "fullness" that's associated with LPs. "Some people mistake this defect for a virtue," Lipshitz says.

173
IS MY MP3 PLAYER A CONCERT-QUALITY DEVICE?

▶ RINGS HOLLOW

Sorry, no. MP3s are compressed files that cut as much as 90 percent of the sound from the original recording by using computer models of human hearing and removing subtle sounds that most of us don't realize we're missing. A compressed recording of a French horn, for example, might lack the slight reverberations from the concert hall.

Instead of filling his digital music player with thousands of songs of crummy sound quality, Grammy Award–winning producer Jim Anderson keeps his iPod stocked with just 55 songs in an uncompressed format, including jazz pianist Keith Jarrett's epic live solo concerts in Germany. (Anderson prefers the lossless AIFF format, in which one minute of stereo audio occupies 10 megabytes.) "If I were to cut the CD down to an MP3, I'd be throwing out all the stuff in the room that makes the piano sound as full as it does," says Anderson, who chaired the Clive Davis Department of Recorded Music at New York University. "I hear the piano exactly as it was at the concert."

174

WHY DOES MY VOICE SOUND SO DIFFERENT WHEN I HEAR IT PLAYED BACK FROM A RECORDING?

▶ **THERE GOES YOUR VOICE-OVER CAREER**
It sounds different because it is different. "When you speak, the vocal folds in your throat vibrate, which causes your skin, skull, and oral cavities to also vibrate, and we perceive this as sound," explains Ben Hornsby, a professor of audiology at Vanderbilt University. The vibrations mix with the sound waves traveling from your mouth to your eardrum, giving your voice a quality—generally a deeper, more dignified sound—that no one else hears.

Through a loudspeaker or recording device, you pick up sound only through air conduction. "The sound we're used to hearing has a lower frequency from the bone vibrations," Hornsby says. "We like that because it sounds rich and full." Many of us cringe at the playback sound because our brain struggles to accept that this foreign voice is our own.

175
WHY DOES ORANGE JUICE TASTE SO BAD
AFTER I BRUSH MY TEETH?

▶ **AWFUL APPETIZER**

Taking a swig of orange juice soon after using toothpaste can fill your mouth with a taste so disgusting that you'd swear you just licked an anvil. But toothpaste doesn't ruin the flavor of all foods and drinks, so what makes the O.J. combo so unpalatable?

Scientists have identified the culprit as sodium lauryl sulfate, a foaming detergent found in most toothpastes. Each taste cell has an outer membrane that contains flavor receptors. The detergent temporarily collapses the membranes and disrupts some of the receptors, explains Linda Bartoshuk, a professor at the University of Florida College of Dentistry. This is what skews our sense of taste after brushing our teeth.

"There are three things you normally detect in the taste of orange juice: sour, sweet, and a note of bitterness," says physiologist John DeSimone of Virginia Commonwealth University, who collaborated with Bartoshuk a decade ago on the definitive study on orange juice and toothpaste interaction. Sodium lauryl sulfate appears particularly adept at dulling the receptors for sweetness and thus blocking the taste of fructose, the sugar in orange juice.

For some reason, toothpaste doesn't interfere with the taste buds that detect sour and bitter flavors. Usually, the citric acid elicits a slight sour flavor, but without tasting the fructose, the sour is enhanced, and the acid's strong bitterness becomes shockingly obvious.

To date, researchers haven't explored what's happening at the cellular level, but DeSimone thinks there's no real scientific incentive for going much further: "You just need to remember to drink your orange juice before brushing."

WHY DOES COFFEE, WHICH SMELLS SO GOOD, MAKE MY BREATH STINK? 176

BACTERIAL BREAKFAST

Coffee, it turns out, transforms your mouth into the ideal breeding ground for pungent bacteria. Like other acidic beverages, such as alcohol and lemonade, coffee dries out your mouth. With less antibacterial saliva to keep the bacteria in check, they reproduce willy-nilly. As metabolic by-products, these bacteria emit hydrogen sulfide, which is the main cause of halitosis.

You're only making matters worse if you take your coffee with milk and sugar. Bacteria love eating both, and sugar also feeds plaque-forming (yet non-stinky) bacteria under which the malodorous bacteria hide.

177 HOW DO I FRESHEN UP MY NASTY COFFEE BREATH?

DIRTY MOUTH

If giving up your morning mochaccino is too awful to contemplate, then try rinsing your mouth with water after you finish your coffee, suggests Harold Katz, founder of the California Breath Clinics. "Saliva does this naturally," he says. "But eating an apple or some celery is also good—they are rough foods with lots of water, which is good for cleaning your mouth."

Cinnamon gum is another effective breath freshener, but not just because its spicy scent masks the stench. Microbiologist Christine Wu of the University of Illinois at Chicago College of Dentistry found that Wrigley's Big Red chewing gum, which contains a negligible amount of sugar, kills up to 50 percent of mouth bacteria, thanks to the antibacterial properties of cinnamon oil and other natural flavor oils.

178

CAN **SOMEONE** USE MY **GPS RECEIVER** TO TRACK ME?

▶ THE FUGITIVE

Conspiracy theorists can rest easy. Handheld global positioning system (GPS) receivers are just that—receivers. They are passive devices that intercept radio signals sent out by a network of 24 primary satellites in medium-Earth orbit. By timing how long a signal takes to reach the device from each of four satellites, a GPS-enabled device can pinpoint the user's position. For someone to stalk you, your GPS device would have to have some way of transmitting that position data, whether by a GPS-enabled cell phone, a radio transmitter, emergency 911 services, or wireless Internet. Most simple handheld GPS devices are passive receivers and have nothing of the kind.

If a GPS device were to be used to track a person without his or her knowledge, it could cross into tricky legal territory. "The idea that someone is able to gather a really nicely aggregated picture of your daily routine is something most people would see as an invasion of privacy," says Lee Tien, a senior staff attorney at the Electronic Frontier Foundation, a nonprofit digital-rights group. "It's not much different from having someone following you around and keeping track of where you are at all times." If you're concerned, there's an easy solution: Just disable the GPS features on your cell phone when you're not using them.

HOW DO QUARTZ WATCHES TELL TIME SO ACCURATELY?

ROCK AROUND THE CLOCK

Watches filled with intricate, tiny cogs and gears are, for the most part, a thing of the past. Modern watches are far simpler as a result of a curious property of quartz called piezoelectricity: When jolted with an electrical current, a quartz crystal vibrates at a very specific frequency.

Quartz has been used in larger, laboratory clocks since 1927, but it was not commercially available in watches until the second half of the twentieth century. "After World War II, watchmakers realized that they had basically hit the limit [of accuracy] with mechanical watches," says Carlene Stephens, curator of the division of work and industry at the Smithsonian National Museum of American History in Washington, D.C.

The big challenge for watchmakers was to create a quartz crystal that was small yet vibrated at a relatively low frequency, because interpreting high-frequency vibrations hogs battery power. Researchers chose 32,768 hertz because this equals 2^{15} vibrations per second, a number that is easily converted using a digital counter.

To get the whole thing going, electricity from the watch's battery induces the quartz to vibrate. These very regular vibrations are then "counted" by the internal circuitry and converted into a signal that, depending on the type of timepiece, drives either the second hand of an analog watch or the counting circuit in a digital watch.

180 IS IT SAFE TO WALK BAREFOOT IN NEW YORK CITY?

WATCH YOUR STEP

The obvious concern is that city sidewalks are dirty. And that's a valid concern, says Daniel Howell, a biologist at Liberty University in Virginia who has lived mainly shoeless for the past few years. "There's a lot of soot, so your feet get blacker than if you were in the woods," explains Howell, who has toured New York barefoot several times. But, he says, a little soot isn't bad.

Surprisingly, germs aren't much of an issue. Disease-causing bacteria are on sidewalks, but getting them on your feet isn't the same as getting infected. "Bacteria typically get into your body through a wet opening, like the eyes or mouth or a cut," Howell says. "If you have an open wound on your foot, keep it bandaged. Or wear a shoe. Shoes are tools, and you should use them when needed."

The key is looking where you put your feet, whether or not they're in shoes. Pete Fernandez is a construction worker in Chicago who has spent the majority of the past four years barefoot. "I've stepped on three nails on job sites, twice with boots and once without," he says. "Hurts either way."

Wounds to a naked foot, however, might be less severe. The barefoot individuals we spoke with reported that they usually feel a sharp object, such as a nail, early enough to pull their foot back before it can penetrate too deeply; shod folks come down on such objects with all their weight because they can't feel it

pierce their shoe's sole. And studies confirm that a shod person's wound is also more likely to become infected, because the nail passes through the person's bacteria-filled shoe and sock and carries those infectious agents into the wound.

As with other barefoot enthusiasts, the soles of Fernandez's feet have developed a supple, leathery pad that lets him walk over most pointy objects, including glass. "As long as it's not sticking straight up, I'm fine," he says. "I get more cuts on my hands. I should probably wear gloves more often."

One of the biggest hindrances to enjoying a barefoot stroll in New York is dog poop. Even though there's no health threat associated with stepping in it, many of our experts agreed on this point. Cody Lundin, who has gone barefoot for 20 years and teaches outdoor survival skills, says, "Let me tell you this: People started going barefoot long before dog [poop] was in cities. I'd rather walk through a cactus plot—at least you see the cacti and know where the needles might be and you can avoid them. Nature is consistent. Dog [poop] can be anywhere, because there are no warning signs in a city. It's all over the sidewalk. There's the hard stuff and the squishy stuff, but it's all horrible to step in." Thanks. Anything else? "Vomit's nasty, too. But it's not fun to step in [poop]."

ARE THOSE PAPER TOILET-SEAT COVERS REALLY PROTECTING ME FROM ANYTHING?

STAY DRY

The paper toilet-seat cover can be a guardian angel for the backside, but only if the seat is dry to begin with. When the cover is set down on a seat that's wet, it ferries bacteria and viruses from the toilet seat up to your bare skin.

The good news is that you're unlikely to contract a disease merely by sitting on a pathogen-covered toilet. Neither viruses like influenza nor the bacteria responsible for illnesses such as strep throat are dangerous unless they come in contact with the mucus membranes—something easily prevented by washing dirty hands with soap before touching your mouth or eyes (which you do already, right?). Most sexually transmitted diseases cannot survive once exposed to air (exceptions are the herpes virus, which can live for a few hours, and hepatitis B, which can linger for seven days). To catch a disease, the seated party would have to have some sort of break in the skin to allow the virus to enter. So if your bum is flawless and you don't mind the yuck factor, go ahead and take a seat.

SHOULD I WORRY AT ALL ABOUT RESTROOM GERMS?

EVERYDAY BIOHAZARDS

Indeed you should—just not about those much-maligned toilet seat–dwellers. The fact is, the germs hiding on the throne aren't the real bad guys. The top sides of toilet seats are low in bacterial numbers compared with surfaces that you actually touch in a public restroom, like the faucet and countertop. And it's not just the bathroom you need to worry about. British researchers have found that computer keyboards are often several times more contaminated with bacteria than the dreaded toilet seat. Says University of Arizona microbiologist Chuck Gerba, "Toilet seats have been getting a bad rap."

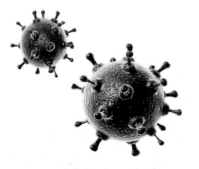

183 WHICH CROP IS BETTER FOR MAKING ETHANOL: CORN OR SUGARCANE?

BOTH ARE TASTY

At a glance, corn appears to be the more energetic crop. By processing 1 ton (907 kg) of corn, producers can make 100 gallons (378.5 liters) of ethanol, whereas 1 ton (907 kg) of sugarcane yields a seemingly paltry 20 gallons (75.7 liters). But sugarcane grows in more tightly packed clumps than corn, so 1 acre (4,046 m²) of sugarcane can produce at least 590 gallons (2,233 liters) of ethanol, compared to 400 gallons (1,514 liters) of ethanol from 1 acre (4,046 m²) of corn.

Converting sugarcane to ethanol is also more environmentally friendly, says José Goldemberg, a Brazilian physicist and secretary for the environment for the state of São Paulo. Distillers in Brazil, the world's largest producer of ethanol, get 8 gallons (30.3 liters) of sugarcane ethanol from just 1 gallon (3.8 liters) of fossil fuel. Corn, the primary source of ethanol in the United States (which is the world's second-largest ethanol producer), yields only 2 gallons (7.5 liters) of ethanol per gallon (3.8 liters) of fossil fuel.

184 SO WHY DOES THE UNITED STATES PREFER CORN?

CORN-SERVATION

The answer is economic and political. Brazil has the right climate for sugarcane and lots of cheap labor. In the U.S., however, government subsidies encourage American farmers to grow more corn, and ethanol demand isn't as high in the U.S. as it is in Brazil, so American farmers make more money growing cane for sugar rather than fuel.

Still, even converting the entire U.S. corn crop into ethanol would satisfy only one-fifth of our fuel needs and leave no corn for animal feedstock and barbeques. A better solution, says Gregory McRae, a professor of chemical engineering at the Massachusetts Institute of Technology, could come from cellulosic conversion technology, which uses enzymes or other chemicals to break down the tough plant fibers in corn husks, grasses, and even trees to unlock the additional energy stored within. Scientists have yet to identify enzymes that can perform this task cheaply and efficiently, though. "It's pretty easy to eat the corn," McRae says, "but chewing through the cob is tough."

IS PREMIUM GAS WORTH THE PRICE? 185

GOURMET GAS

With gas prices seemingly always on the rise, some drivers may find themselves hesitating at the pump, wondering what they get for splurging. And rightfully so: For most engines, the only premium feature of high-octane gas that drivers will experience is its price tag. Unless your vehicle is one of the 10 percent of those on the road that require premium—typically sports and luxury models—there is absolutely no benefit to premium gas.

High-octane gas helps prevent engine "knock," the pinging sound that occurs when the air-fuel mix pre-ignites under pressure in a cylinder before being lit by a spark plug. Pre-ignition can damage a piston by propelling it down before it completes its upstroke. The higher a fuel's octane rating, the less likely it is to ignite from pressure alone.

Most cars built since the mid-1980s, however, have computers that prevent knock by constantly adjusting the timing of ignition. So why do some cars still need premium? The more an engine pressurizes the fuel before igniting it, the more power it extracts from the explosion. High-performance cars, such as Mercedes and Porsches, are designed to maximize power in this way, so they need a gas that's not prone to pre-ignite. Your station wagon wasn't built to compress gas as densely, though. It does just fine on regular.

186 CAN I USE CHEAP GASOLINE IN MY EXPENSIVE CAR?

FUEL ECONOMY

If you drive a car with a high-performance engine, you should stick to the required premium or you risk damaging it, says Steve Mazor, chief automotive engineer for the Automobile Club of Southern California. Although other fine-tuned cars in which premium is only "recommended" will run on regular fuel, the constant adjusting comes at the cost of horsepower. Fill these cars with regular, and you might notice a dip in acceleration or a drop in gas mileage. Bottom line: read your owner's manual and follow the directions. "If your owner's manual says to use regular gas," Mazor says, "putting premium in your car is just wasting money."

187

I PAID EXTRA TO MAKE MY CROSS-COUNTRY FLIGHT "CARBON-NEUTRAL." AM I SAVING THE PLANET, OR AM I A SUCKER?

▶ GREEN BANDWAGON

In the past few years, the quest for what's known as carbon neutrality—essentially, canceling out the greenhouse-gas emissions caused by your activities by supporting clean-air projects—has become a multimillion-dollar industry. Corporations, big-budget movies, rock bands like the Rolling Stones and Coldplay, and even the Super Bowl are now making carbon neutrality part of their brand identity, offsetting their emissions by funding such projects as building wind farms in Nebraska, providing solar panels to Africa, and planting mango trees in India.

Dozens of companies will gladly take your money to help assuage your global-warming guilt. For example, the online travel services Expedia and Travelocity offer customers the option of canceling out the CO_2 resulting from their travel. According to the Environmental Protection Agency, the average American generates 10 tons (9,072 kg) of carbon dioxide yearly, which most companies estimate can be offset for $100 to $200.

188   SO A LOT OF PEOPLE ARE DOING IT, BUT DOES THAT MEAN IT'S WORTHWHILE?

▶ CAVEAT EMPTOR

Is it all a scam? Not necessarily. The key is to make sure your money is funding projects that actually do prevent emissions and that reduce them beyond what would be achieved by business as usual, a concept known in the industry as "additionality." One organization that works to educate consumers on which programs measure up is the nonprofit Clean Air–Cool Planet (cleanair-coolplanet.org). They produce an evaluation of offset providers that consumers can download, rating each on a scale of 1 to 10. Of thirty companies evaluated, only eight earned a 5 or higher.

Meanwhile, the European watchdog group Gold Standard Foundation (cdmgoldstandard.org) certifies well-run, legitimate carbon-offsetting projects using a system that works much like the "fair trade" coffee and "100% organic" labels. "Beware of any old company claiming to be carbon-neutral," says former marketing director Jasmine Hyman. "It's the wild, wild West."

The takeaway? Instead of paying an intermediary to offset your emissions, do a little research and make a contribution directly to a well-reviewed organization (Atmosfair and NativeEnergy are two good bets). And remember, the best way to guarantee that your personal emissions will be reduced is to reduce them yourself. Money's not enough, something Coldplay found out the hard way. The mango trees in India that they funded died due to a lack of maintenance.

189 WHY DO THE COLORADO ROCKIES KEEP THEIR BASEBALLS IN A HUMIDOR?

SMOKIN' FASTBALLS

Tune into a Colorado Rockies baseball game, and you're bound to hear one of the announcers mention the team's most well-known piece of lore: They store their baseballs in a humidor. Wait, you may be asking yourself, aren't humidors used for cigars? Indeed, cigar aficionados keep their cigars in a humidity-controlled environment to prevent the tobacco leaves from drying out, which would affect their flavor. The Rockies aren't worried about the flavor of those baseballs, but rather about dried-out balls carrying farther and driving up scores. Why does this matter? Because Coors Field was well on its way to developing a reputation as a park that was seriously unfriendly to pitchers—and very friendly to home runs.

From the 1995 to 2001 seasons, National League pitchers at Coors Field recorded a horrendous earned run average (ERA) of 6.50, more than two runs a game higher than the 4.37 ERA recorded at other stadiums.

Fans and the media attributed the numbers to Denver's mile-high thin air. But in the winter of 2002, based on a hunch that the balls might be drying out and losing weight in Denver's arid climate, engineers at Coors Field installed a humidor for storing game balls. Since then, N.L. pitchers have posted a 5.46 ERA at Coors. But scientists still can't say exactly why it's so effective.

According to a 2004 study by physicist David Kagan of California State University at Chico, keeping the balls at 50 percent relative humidity lowers their coefficient of restitution, a.k.a. bounciness. This means that humidified balls don't bounce off the bat as powerfully as dried-out ones do, making for a less batter-friendly pitch. Edmund Meyer and John Bohn, physics professors at the University of Colorado, later found that the added moisture does not change a ball's size and shape—and thus, its aerodynamics—which seems to verify Kagan's explanation for the humidor's success.

190 DO PITCHERS KNOW MORE THAN PHYSICISTS?

BASEBALL UNIVERSITY

It very well may be that Rockies pitchers discovered something scientists hadn't realized. Former Rockies pitcher Shawn Chacon griped that pre-humidor balls were as slippery as pool balls, making it difficult to impart enough spin to execute breaking pitches. Lloyd Smith, a professor of mechanical engineering at Washington State University who tests bat materials, and physics professor Alan Nathan of the University of Illinois have investigated this effect and have qualitatively corroborated physicist David Kagan's findings. "We're fairly confident that the effects are small," Smith says. "But in baseball, even small effects can be important." Just ask the Rockies pitchers.

181

DO PEANUT SHELLS HAVE ANY COMMERCIAL VALUE?

▶ BLEACHER TREASURES

Believe it or not, there is a niche market for those woody hulls, although the empty shells you toss away at ballgames are by no means a gold mine. Until two or three decades ago, shells were discarded or burned, says Bob Parker, vice president of Georgia-based Golden Peanut Company, one of the largest peanut shellers in the country. (Snack companies like Planters, it turns out, don't shell their own nuts, instead buying them from outfits like Golden Peanut.) But pollution concerns and disposal costs forced companies to find new ways to reuse the nation's mountain of empty shells, which last year totaled 375,000 tons (340 million kg).

SHELL COLLECTING

Typically, the husks are ground up and sold as a source of bulk fiber for livestock feed or as an absorbent floor covering for chicken houses. Some of Golden Peanut's hulls are ground into a powder and mixed with insecticide or sold as filler for bricks. Others are shipped to Europe and burned as biomass for energy, an idea that scientists are exploring in the U.S. as well. Golden Peanut won't reveal how much it hauls in for the shells, but compared with the nutmeat they shelter, the husks' commercial value is, well, peanuts. Selling them off barely covers the cost of otherwise disposing of them.

192

CAN THE HUMAN BODY BE MADE RADIATION-RESISTANT?

Radiation can wreak havoc on healthy cells during cancer treatments, even as it works to destroy cancerous cells. To combat this effect, a team at the Albert Einstein College of Medicine of Yeshiva University has developed a technique that may protect patients' bodies from radiation from the inside out. By coating tiny silica particles in melanin, a pigment that fights DNA-damaging free radicals, and injecting these nanoparticles into mice, researchers were able to protect the mice from some of the dangers of radiation exposure. If the procedure proves to be successful in human subjects, the injections could be used to protect astronauts and other people who routinely face radiation on the job.

193

HOW CAN NANOTECHNOLOGY IMPROVE MY SHOES?

The must-have sneakers of 2050 may well have soles made from viscoelastic super-rubber. This extremely flexible, strong, and temperature-resistant material was created by Japanese researchers from a network of interconnected nanotubes. The intertwining links between the nanotubes lend the super-rubber its ability to maintain its shape under extreme temperatures, as well as its amazing malleability. While far too expensive to manufacture widely today, such materials could be the building blocks of tomorrow's fashions, factories, and spacecraft.

▶ CAN GPS WORK UNDERGROUND?

Since GPS technology relies on satellite signals, it's hard to avoid losing your way as soon as you drive into a tunnel. But GPS isn't the last word in navigation. Israeli researchers have created tiny optical gyroscopes the size of a grain of sand that could solve this problem. Optical gyroscopes track your movement without needing outside reference points. Such technology is nothing new, but traditionally these machines have been way too large and heavy to fit inside your cell phone. With these new nanogyros, it could become far easier to use portable devices to keep track of your location inside or underground.

195

HOW STRONG IS THE WORLD'S STRONGEST PAPER?

Really strong—if you can actually call it paper. Researchers have created a flexible, absorbent, magnetic "paper" from a cellulose-based aerogel capable of withstanding 400,000 pounds per square inch (28,100 kg/cm²) of pressure. Aerogels are often used as insulators, and they generally combine significant strength with low density. To make this version, researchers soaked cellulose in metallic compounds, which gave the material some metallic properties. The result is an unusually flexible aerogel (it's foldable, which is unusual for similar materials) that can be manufactured from cheap plant materials. While it's unlikely that anyone would actually replace normal paper with this wonder-paper anytime soon, it's likely to have a wide variety of uses in industrial production.

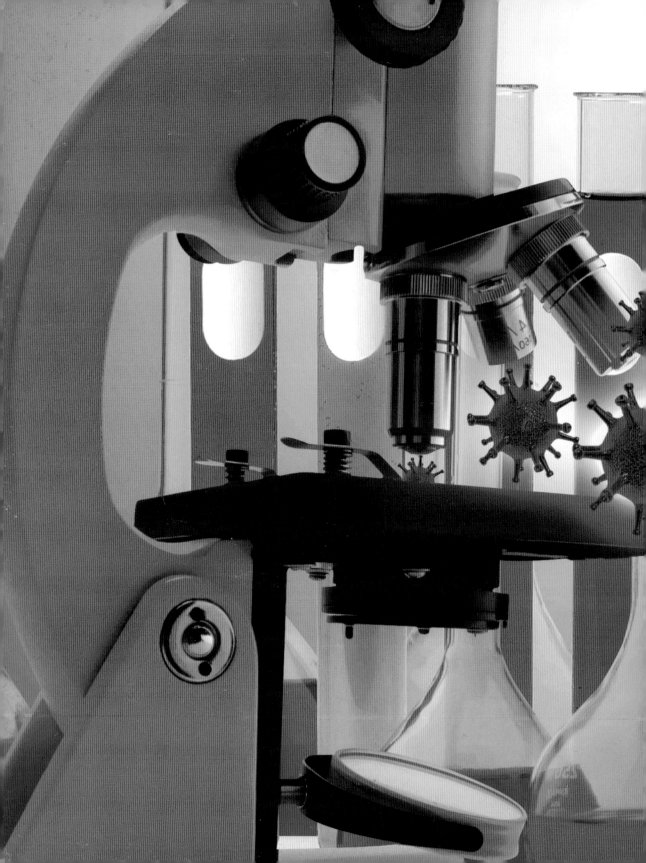

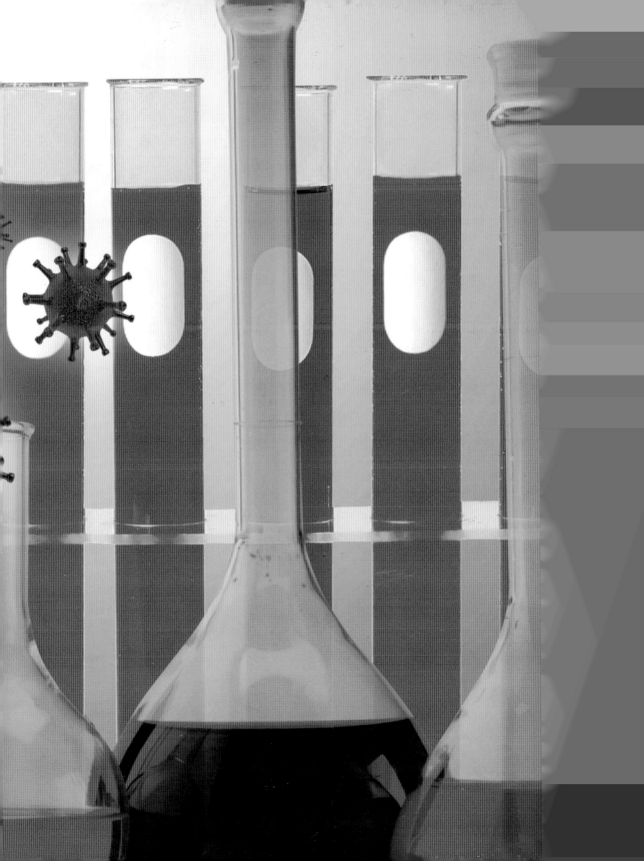

196
CAN A PAPER CUT BE DEADLY?

▶ DEATH BY A THOUSAND TINY CUTS

A person with normal clotting ability would have to lose nearly 40 percent of his blood almost immediately to die of blood loss. The arteries with this bloodletting potential, though, such as the radial artery in the arm and femoral artery in the leg, are buried under too much flesh to be nicked by even the heaviest manila cardstock. "It would be tough to kill yourself on a paper cut," says Beau Mitchell, a bleeding-disorder specialist at the New York Blood Center, an organization that supplies blood to hospitals.

A stationery slice could turn deadly, however, for the 12,600 people in the U.S. with severe hemophilia and the 200 Americans with a disorder called Glanzmann's thrombasthenia. If one of these people sliced an exposed blood vessel, like the one under the tongue, their blood would not be able to clot to plug the wound. Glanzmann's patients are especially vulnerable, Mitchell says, and could lose 25 percent of their blood within eight hours from such a cut. Without medical treatment, their bodies couldn't produce enough new blood cells to replenish those lost, and they would die within a few days.

Although people with these diseases should probably avoid licking envelope seals, we should all avoid ninjas armed with paper daggers. According to Ronald Duncan, a master of the martial art ninjutsu, anyone can fold a piece of paper, origami-style, to fashion a sharp knife. Duncan trains police officers and the military to look out for these weapons because a jab to the carotid artery in the neck could be fatal. "A few other parts of the body can bleed out in 35 seconds if someone is really adept," he says. "But we try not to make this information available to too many people."

ALCOHOL
96°

WHY DOES BLOOD MAKE SOME PEOPLE SQUEAMISH BUT NOT OTHERS?

▷ BAD BLOOD

Looking at blood can be hard on anyone, but for some people, it can be a huge problem. Up to 30 percent of children are afraid of the sight of blood, a response that usually continues into adulthood, according to the definitive study on the topic by Isaac Marks of the Institute of Psychiatry in London. That study also revealed that approximately 15 percent of the adult population faints when donating blood.

Fainting at the sight of blood generally comes from an overactive vasovagal response, an evolutionary fear reflex. This response slows down your heart rate and lowers your blood pressure, causing blood to drain to your legs. This means that less oxygen-rich blood is going to your brain, which is what causes people to feel light-headed or even pass out.

This survival mechanism is nice if, say, you need to play dead in the presence of a predator, which may be the origin of the response. And if you're bleeding, the reduced heart rate might help prevent too much blood loss. But in most situations—especially emergencies when you need to be, you know, awake—it's a nuisance.

The more a person is exposed to the thought or sight of blood, however, the less bothersome the phobia becomes, says Alan Manevitz, a psychiatrist at Weill Cornell Medical Center in New York. This could explain why experienced surgeons deal so well with blood. This type of "systematic exposure" is a common treatment for specific phobias, and in this case, Marks suggests, it could save lives: Ill people who have overcome their fear of having blood drawn are more likely to seek medical care.

WHAT AWESOME THINGS CAN BACTERIA DO?

CLEAN UP OIL SPILLS

Following the enormous Deepwater Horizon oil spill in the Gulf of Mexico, bacteria played a surprisingly heroic role. In June of 2010, scientists studying the spill found that methane was concentrated in the water at levels up to 100,000 times above normal levels, and it looked likely that the buildup would take years to dissipate. Within four months of the spill, however, scientists discovered that bacteria had consumed more than 200,000 metric tons of methane, and the levels had gone back to normal. The adaptability of bacteria (which can be pretty frightening from a disease-fighting standpoint) allowed methane-eating strains to rapidly proliferate in the newly methane-rich environment, helping to repair the Gulf in the process.

ACT AS LIVING FUEL CELLS

Researchers at Penn State have created a fuel cell that uses bacteria to simultaneously create power, purify wastewater, and desalinate seawater, a process that usually consumes large amounts of energy. The elegant setup relies on bacteria that naturally oxidize wastewater, setting off an exchange of charged ions that produces power and removes the salt from seawater poured into a central chamber. Although the fuel cells aren't optimized yet and won't be useful on an industrial scale for some time, this could be a key technology for water-starved areas of the globe.

BUILD TINY PYRAMIDS

Why build tiny machines to do your bidding when you can use tiny organisms instead? Researchers at the NanoRobotics Laboratory of the École Polytechnique de Montréal have used a magnetic field to assemble a swarm of bacteria, then manipulated the magnetic signals to direct the swarm to assemble a tiny pyramid from microscopic blocks. As this technique is perfected, such bacteria could carry out the functions of nanobots—administering targeted medications in a patient's bloodstream or assembling more useful nanostructures than tiny pyramids, for example.

STORE INFORMATION

Your hard drive may be made obsolete by the same organisms that just gave you food poisoning. Chinese University students in Hong Kong have developed a system that uses the DNA of *E. coli* bacteria to encode and store information—in theory, a gram of the bacteria could store as much data as 450 two-terabyte hard drives. The students have shown how it would be possible to store your pictures, songs, and even videos within cells, using system maps for easy retrieval. And since bacteria keep replicating, the system could potentially preserve data for millennia, allowing your embarrassing Facebook posts and photos to outlive you indefinitely.

199
WILL DRINKING CARBONATED BEVERAGES WEAKEN MY BONES?

▶ **BREAKING BAD**

Maybe—but only if you're drinking several gallons of seltzer a day. Here's the chemistry that has soda drinkers worried: As carbon dioxide hits the water in your blood, it turns into carbonic acid. Too much acid in the blood can lead to a condition called acidosis, which could intercept small amounts of calcium from food as it makes its way to your bones—or steal it from them directly. Your greater concern, though, says endocrinologist Robert Heaney of Creighton University, should be the vomiting, headaches, and impaired organ function that result from extreme acidosis.

The acid content in a carbonated beverage is 5 to 10 percent of what the body's metabolism naturally produces, Heaney has found, which is far too little to interrupt the calcium absorption of bones. In general, he says, the carbonation in soda has no ill effect on bone-mineral content.

Other ingredients in soda might rob a small amount of calcium from bones. Caffeine causes the kidneys to pull sodium from the blood using proteins that accidentally scoop up calcium ions as well. The body reverses this effect within 24 hours, however. Another commonly cited culprit is phosphoric acid, an ingredient in colas. Studies have indicated that if the ratio of phosphorus to calcium in your body tips too far toward phosphorus, it can cause bone loss over time. So although a can of pop once in a while won't a brittle bone make, Heaney urges drinking a tall glass of milk to keep your bones good and strong.

200

CAN SUCKING IN MY GUT GIVE ME SIX-PACK ABS?

▶ TRY A CORSET

Sucking in your gut is a good quick fix for hiding a flabby physique, but it won't turn your midsection into a washboard anytime soon.

Sucking in your stomach *is* exercise, though, says Priscilla Clarkson, a professor of kinesiology at the University of Massachusetts. "Depending on how many times a day you did this, you can actually tone the muscle," she says. But, she warns, this practice isn't an effective way to build muscle or burn fat.

TONING TIP

If you still want to give it a go, professional trainer Christian Finn suggests trying the "stomach vacuum," a gut-sucking exercise that competitive bodybuilders use. To do it properly, start by taking a deep breath and lifting your chest. Next, exhale and quickly suck in your stomach as far as you can and hold for three seconds.

Start with 10 repetitions every other day, increasing the length of each rep as you improve. "The benefit of the vacuum is that it takes very little time to perform, and you can do it anywhere," Finn says. "Even lying in bed." But, he adds, it mainly flexes internal muscles, so you'll see results only if you already have little fat around your belly. Maybe it's better to hit the gym after all.

WHAT WOULD HAPPEN
IF I ATE A TEASPOONFUL OF
WHITE DWARF STAR?

PLEASE DON'T

"Everything about it would be bad," says Mark Hammergren, an astronomer at Adler Planetarium in Chicago, beginning with your attempt to scoop it up. Despite the fact that white dwarfs are fairly common throughout the universe, the nearest is 8.6 light-years away. Let's assume, though, that you've spent 8.6 years in your light-speed car and that the radiation and heat emanating from the star didn't kill you on your approach. White dwarfs are extremely dense stars, and their surface gravity is about 100,000 times as strong as Earth's. "You'd have to get your sample—which would be very hard to carve out—without falling onto the star and getting flattened into a plasma," Hammergren says. "And even then, the high pressure would cause the hydrogen atoms in your body to fuse into helium." (This type of reaction, by the way, is what triggers a hydrogen bomb.)

Then you'd have to worry about confinement. Freeing the sample from its superdense, high-pressure home and bringing it to Earth's relatively low-pressure environment would cause it to expand explosively without proper containment. But if it didn't blow up in your face—or vaporize your face, since the stuff's temperature ranges between 10,000°F (5,500°C) and 100,000°F (55,500°C)—and you somehow got it to your kitchen table, you'd be hard-pressed to feed yourself: A single teaspoon would weigh in excess of 5 tons (4,500 kg). "You'd pop it into your mouth and it would fall unimpeded through your body, carve a channel through your gut, come out through your nether regions, and burrow a hole toward the center of the Earth," Hammergren says. "The good news is that it's not quite dense enough to have a strong enough gravitational field to rip you apart from the inside out."

BUT WHAT WOULD A
STAR TASTE LIKE?

MOONBEAMS AND UNICORNS

Eating a star probably wouldn't be worth the trouble anyway, Mark Hammergren laments. White dwarfs are mostly helium or carbon, so your teaspoonful would taste like a whiff of flavorless helium gas or a lick of coal. But if you're desperate for a taste of star, you don't really need to travel 8.6 light-years—your fridge is full of the stuff. Most of the elements that make up our bodies and everything around us were formed in the cores of stars and then belched out into the universe over billions of years. Basically everything you eat has at some point been part of a star. Might we recommend a nice piece of star fruit?

203

WHY DO PEOPLE SOMETIMES CALL SUNLIGHT "VITAMIN D"?

▶ **SOAK UP THOSE RAYS**
You need activated vitamin D3, also known as calcitriol, for the maintenance of proper blood chemistry and healthy bone tissue. You can get the vitamin from supplements, but it's cheaper (not to mention more pleasant) to make it the old-fashioned way—by going outside and soaking up some rays. The biological process is pretty basic: A compound in the skin reacts with ultraviolet light from the sun to produce vitamin D precursor molecules. Two carbon atoms in the previtamin then spontaneously rearrange to create the bone-bulking vitamin D3. This all happens quickly; after just a few minutes in the sun, the body has absorbed enough light to produce several times the necessary daily intake of the vitamin. And what if you don't get enough vitamin D? If you're a kid, rickets. If you're an adult, adult rickets. So, if you're feeling brittle, bust out the sunglasses.

204
WHY DOES SUNLIGHT MAKE SOME PEOPLE SNEEZE?

▶ **GESUNDHEIT**

When Michael Stephani steps out into the bright sunshine after spending a couple hours in a dark movie theater, a brief sneezing fit ensues. Sun sneezes are typical for Stephani, whose oldest son also sneezes at the sun. "It doesn't annoy me at all," he says. "I think it's rather energizing, welcoming the sun with a sneeze."

The sun induces sneezing in 10 percent of the U.S. population, says Louis J. Ptácek, a neurologist at the Howard Hughes Medical Institute in Maryland and a professor at the University of California at San Francisco. Just how and why this happens, though, has remained a mystery ever since Aristotle penned the question some 2,300 years ago.

Research suggests that the photic sneeze reflex, or PSR, is inherited, but scientists have yet to pinpoint the gene or genes responsible. "There's precious little known about PSR, and part of that is because it's not a disease," Ptácek says. "No one dies from it."

One theory is that the gene involved—whatever it is—crosses wires in the brains of those with PSR. For these people, when light enters their eyes, it activates their brain's visual cortex but also stimulates the motor region that causes the diaphragm to quickly contract, forcing a sharp burst of air out through the nose.

Although sun-triggered sneezing is more of a quirk than a serious condition, Ptácek says, understanding the science behind it could shed light on the underlying biology of other "reflex" phenomena, such as certain types of epilepsy.

205 CAN "BRAIN FREEZE" CAUSE LONG-TERM BRAIN DAMAGE?

CRYOGENIC CRANIUM

We've all sucked down a milk shake so quickly that it causes a sudden headache—the dreaded brain freeze. You know you should take a break, but . . . milk shake. So tasty. Must. Drink. Could chugging the rest of that shake cause lasting brain damage?

First, let's get the name straight. "This condition is referred to as an 'ice-cream headache,'" says Stacey Gray, a sinus surgeon at the Massachusetts Eye and Ear Infirmary in Boston. "It's a very technical term." Although there's no published paper saying as much,

a milk shake slurped too quickly probably does not actually lower brain temperature. Besides, Gray says, the temporary pain can't do any harm because it has nothing to do with the brain.

When you feel that characteristic stabbing pain, the cold drink has likely touched off a branch of the trigeminal nerve in your mouth, triggering pain in the nerve that's responsible for facial sensation. It may seem like someone's stabbing you through the frontal lobe, but rest assured, your brain never feels a thing.

206 BUT WHAT WOULD HAPPEN IF YOUR BRAIN REALLY DID GET COLDER?

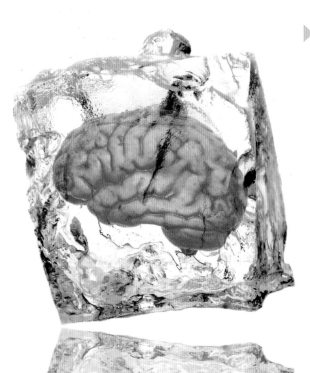

BRAINSICLE

Even if a cold drink did chill your brain a bit, it wouldn't be a big deal. Neurosurgeons including Johns Hopkins Hospital's Rafael Tamargo often take the brain from its cozy resting range of between 98.6°F and 100.4°F (37°C–38°C) all the way down to 64°F (17.8°C). "There are situations, particularly for correcting blood-vessel problems, where we cool the brain in order to stop circulation to an area to perform our work safely," he says. When the brain is chilled to 68°F (20°C), its metabolism and electrical activity drop to 15 percent; surgeons reduce it to 64°F (17.8°C) for good measure.

"Even if the patient wasn't anesthetized, at that temperature they would be in a noninteractive state, unable to sense stimuli or produce a response," Tamargo says. "But once you warm the brain up, it picks right up from where it left off. It's not harmful at all." So whether your brain is frozen or not, if you can handle a little pain, slurp away.

WHY DO I GET A HEADACHE WHEN I EAT ICE CREAM TOO QUICKLY?

SLOW AND STEADY

An ice-cream headache hits when a cold substance makes contact with nerves in the roof of your mouth. A nerve located in the back of your throat (impress and annoy your friends by calling it the sphenopalatine ganglion) stimulates the trigeminal nerve, the largest of the sensory nerves that lead from your face to your brain. The result is that characteristic stabbing pain, centered in the mid-frontal part of the brain. Migraine sufferers are typically quite susceptible to cold-induced headaches, and a 10- or 20-second brain freeze can often be a trigger for a longer migraine attack. There's really no way to avoid such headaches if you insist on wolfing down your icy dessert, says Seymour Diamond, founder and director emeritus of the Diamond Headache Clinic in Chicago. Next time you sip a smoothie, just remember to take it slow.

208

I'VE HEARD THAT CHOCOLATE HAS SOME HEALTH BENEFITS. IS THIS TRUE?

▶ A BROWNIE A DAY . . .

You're in luck! In the past decade, scientists have put forward enough evidence to lend serious support to the claim that chocolate, or at least some of the chemicals therein, is good for you. Chocolate is rich in a class of chemicals known as flavonoids, a type of antioxidant. Scientists believe that these chemicals fuel chocolate's health benefits, but this is tricky: The exact mechanisms by which antioxidants themselves work their magic are not fully understood, but they are known to clean up free radicals—charged atoms, ions, or molecules that damage healthy cells—in the body.

A 2009 study (run by the Nestlé Research Center in Switzerland) fed 30 subjects 1.4 ounces (40 g) of dark chocolate a day for two weeks. That intake, researchers reported, drove a significant reduction in levels of stress-associated hormones, perhaps explaining why a bit of chocolate can help you unwind after a busy day.

The antioxidant effect of chocolate also appears to reduce inflammation, a condition at the root of many diseases. This effect can improve blood flow by making blood platelets less sticky; it also lowers bad cholesterol (LDL) and protects against coronary heart disease. A 2008 study revealed that eating a quarter ounce (7 g) of dark chocolate per day was enough to considerably reduce the levels of C-reactive protein, which is associated with increasing inflammation.

Here's one you'll really like: Chocolate's antioxidants might even improve your skin. In a German study, women who drank a daily beverage fortified with chocolate flavonoids experienced 15 to 25 percent less skin redness in response to exposure to UV light. The hydration and texture of their skin also improved, which the scientists attributed to improved blood flow to the skin.

209 CAN CHOCOLATE MAKE ME SMARTER?

▶ BRAIN FOOD

A mouthful of Hershey's might even make you better at math. In 2009, scientists at Northumbria University in the U.K. found that drinking a hot chocolate beverage containing 500 mg of flavonoids helped people perform better on a simple math problem—counting backward by threes from a random number—than study subjects who did not. OK, that's not multivariable calculus, but the researchers believe that the chocolate drinkers experienced an advantage because chocolate increases blood flow to the cerebral cortex, the part of the brain responsible for handling complex tasks.

Now, before you go hog wild, consider this: The 500 mg of flavonoids in the math study was about the equivalent of mowing down five chocolate bars. You don't need your newfound math skills to run the chocolate bar/waistline equation: For now, eating enough chocolate to get the health benefits might not be worth it.

210
CAN RIDING
ROLLER COASTERS
TOO OFTEN BE BAD FOR MY HEALTH?

▶ UPS AND DOWNS

Not if you're healthy to begin with. The forces generated by a coaster can equal those of a fighter-jet takeoff or shuttle launch, which can prevent blood from reaching the brain and lead to blackouts. But on amusement rides, this force rarely lasts longer than three seconds; even with multiple loop-de-loops, that's safe.

No serious long-term risks have been associated with frequent riding, and most experts agree that your main concerns should be temporarily cramped muscles and head and neck aches. You might experience some irritation from the rides' restraints—lap bars can leave bruises, and shoulder harnesses have been known to cause "nipple burn." And it can take a while for your inner ear to return to normal after the prolonged twisting and turning, so you may feel off-balance for up to a few days, says John Gerard, a San Diego physician and member of the American Coaster Enthusiasts.

Perhaps the best case study is rail junkie Richard Rodriguez, who has been riding roller coasters for hours at a time for 30 years. The only permanent damage he's noticed is stiff, almost arthritic hands, from constantly gripping his harness and bracing for drops and turns. "They don't work as well as they used to," he says. "Sometimes I have to push my right hand open with my left."

We caught up with Rodriguez in 2003, right after he set the Guinness World Record for continuous time on a coaster: 401 straight hours (that's 16 days, 17 hours) set in Blackpool, England. "My balance has been a little off and I've wanted to lie down a lot, but I always bounce back quickly," he reported. "I'm still peeling a bit from the windburn—I look like I spent a week at the beach." All the screaming kids on the coaster did cause him frequent headaches (he's fine now), but at least he managed to escape the nipple burn.

DO CELLS MAKE NOISE?

▶ **KEEP IT DOWN IN THERE**

You have to listen very, very closely, but yes, cells produce a symphony of sounds. Although they won't win a Grammy anytime soon, the various audio blips produced by cells are giving scientists insight into cellular biomechanics and could even be used to help detect cancer.

Researchers at the University of California at Los Angeles studying brewer's yeast discovered that the yeast's cell walls vibrate 1,000 times per second. These motions are too slight and fast to be caught on video, but when converted into sound, they create what the scientists describe as a high-pitched scream. (It's about the same frequency as two octaves above middle C on a piano, but it's not loud enough to hear with the naked ear.)

"I think if you listened to it for too long, you would go mad," says biological physicist Andrew Pelling, now at the University of Ottawa. Pelling and Jim Gimzewski, his advisor at UCLA, theorize that molecular motors that transport proteins around the cell cause the walls to vibrate.

CAN SCIENTISTS LEARN ANYTHING FROM HUMAN CELL NOISE?

Turns out, it's a little harder to get sound out of a human cell than from a yeast cell. So far, scientists have not observed mammalian cells that audibly shimmy on their own, at least in part because animal cells' wiggly membranes are less likely to vibrate than the sturdy cell walls of yeast and plants. But human cells certainly squeal when zapped with light, and this could end up being surprisingly useful for medical science, particularly cancer research.

When Richard Snook and Peter Gardner, biologists at the University of Manchester in England, blasted human prostate cells with infrared light, their microphones picked up thousands of simultaneous notes generated by the cells. Through statistical analysis of these sounds—which are created as the cells rapidly heat up and cool down, causing vibrations in the air molecules directly above them—Snook and Gardner can differentiate between normal and cancerous cells. "The difference between a healthy cell and a cancer cell is like listening to two very large orchestras playing their instruments all at the same time," Gardner says. "But in the cancerous orchestra, the tuba is horribly out of tune."

Gardner is fine-tuning the technique in hopes of replacing current, unreliable pre-biopsy prostate-cancer tests. His ultimate goal is to reduce the number of prostate biopsies performed, 75 percent of which come back negative.

212
CAN PHYSICIANS TREAT ANIMALS?

▶ HORSE PILLS

Physicians and veterinarians agree: If it looks like a duck, walks like a duck, and is sick like a duck, it's best for it to be treated by someone trained to treat a duck. Faced with such a scenario, physicians would be armed only with what they know about human biology. And that doesn't go very far, says Rika Maeshiro, the director of Public Health and Prevention Projects for the Association of American Medical Colleges. If, she explains, a physician simply had to treat an animal, three factors stand in the way of effective care.

First, the diseases animals contract could easily be unfamiliar to a physician, and animals aren't able to tell a doctor what's bothering them. And even if the doctor identified the problem, the same injury to a human can have different consequences for an animal, adds Kimberly May, a veterinarian and spokesperson for the American Veterinary Medical Association. "A broken leg isn't fatal for a person, but very well could be for a horse," she says.

Second, as any doctor knows, some cases call for training in a specific type of medicine. "There's a big difference in treating a wound and treating congestive heart failure," Maeshiro says. In most emergencies, a versatile trauma doc would have the best shot at providing animal care.

Third, some animals might be too exotic. The bodies and systems of mammals like dogs, cats, pigs, and cows have some resemblance to humans', so a doctor would at least have an idea where to start, Maeshiro says. "But if someone comes up to you with a snake . . . "

213 SO CAN VETERINARIANS TREAT HUMANS?

▶ IS THERE A VET IN THE HOUSE?

What if you, a human, collapsed on a plane? Could a veterinarian help? Possibly, because anatomy and common ailments will be familiar. But it depends on what's wrong with you. Vets aren't trained to treat human infectious diseases, Kimberly May says, and there are differences in CPR techniques. "Outside of extreme situations, we don't like physicians providing veterinary medicine, and we vets stick to animals," she adds. "It's just professional courtesy."

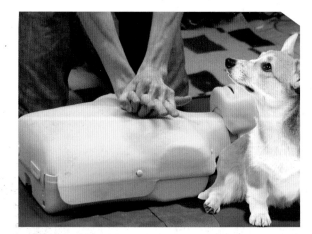

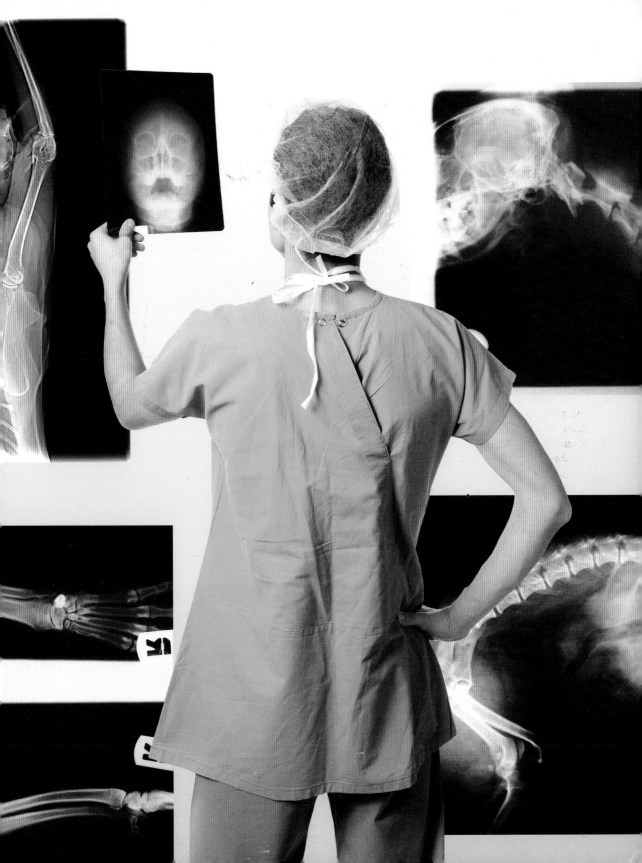

214 WHY ARE SO MANY KIDS ALLERGIC TO PEANUTS?

▶ **PB&J PLAGUE**

The number of school-age kids with peanut allergies has doubled in the past decade. Yet scientists can't quite put their finger on what makes the legume such a threat, or why the allergy has become so prevalent.

Theories abound, though, and most involve an overactive immune system. "We have done such a good job of eliminating the threats that the immune system is supposed to manage that it's looking for something to do," says Anne Muñoz-Furlong, former CEO of the nonprofit Food Allergy and Anaphylaxis Network. Parents feed their kids more ready-made snacks these days, she says, and many of those contain peanuts or their derivatives. "We're bombarding the immune system with these [food-based] allergens, so it's attacking those instead." Indeed, food allergies in general are on the rise.

But peanuts seem to trigger especially violent immune reactions. This might be because they contain several proteins not found in most other foods, posits Robert Wood, an allergy specialist at Johns Hopkins University, and the structure of these proteins can stimulate a strong immune response. Research suggests that roasting peanuts, as American companies do, might alter the proteins' shape, making them an even bigger target. Allergy rates are lower in China, where it's customary to boil peanuts, which damages the proteins less. (It's worth noting, though, that China is also more polluted, so people's immune systems might be concentrating on traditional threats.)

215 IS THERE ANYTHING WE CAN DO TO AVOID PEANUT ALLERGIES?

▶ **GO PLAY OUTSIDE**

Typically, the immune system treats peanuts as safe, but some scientists believe that early and heavy exposure to peanut-laden products might cause it to misidentify them as dangerous. This theory is strengthened by the fact that 8 out of 10 allergic kids have a reaction the first time they eat a peanut, indicating a previous indirect exposure, possibly even in the womb or through breast milk.

Or maybe it's all the video games. Scientists think that vitamin D, which the body needs sunlight to make, helps the immune system label substances as innocuous and thus build up a tolerance. Children who spend less time outdoors tend to be deficient in D, Robert Wood says, so their bodies might mislabel peanut proteins as dangerous. Parents looking to protect their kids might consider sending them outside—and not washing their hands when they come home.

HAIR-BORNE DISEASE

Finding a hair in your spaghetti is gross, no question. But it is not, for the most part, a health threat. It's so benign that the Food and Drug Administration, in its Food Code guidelines, doesn't even place a limit on strands per plate. The FDA has received no reports of people getting ill from ingesting hair found in food.

Here's why you shouldn't worry, explains Maria Colavincenzo, a dermatologist at Northwestern University who specializes in hair: Hair is made up of a densely packed protein called keratin, which is chemically inactive in hair and won't cause any problems if digested. It's possible that staph bacteria, which can upset the stomach and bring on a case of diarrhea, could hitch a ride on a strand. But it's very unlikely, Colavincenzo says, that the tiny amount of staph that can hide on a hair or two is enough to lead to gastrointestinal problems.

The only real scenario in which hair would pose a threat, she continues, is if you ate a whole head's worth. Large quantities of the stuff can do to your digestion what they do to your shower drain. Ingesting that much could make long clumps of hair, called trichobezoars, form in your stomach and cause abdominal pain and other symptoms.

217 SO COULD HAIR ACTUALLY BE EDIBLE?

SPAGHETTI AND HAIRBALLS

Sure, once it's processed. In fact, you might well have eaten hair today. Food manufacturers often use L-cysteine, an amino acid in keratin, to stabilize dough and perk up the taste buds that detect salty, savory flavors. Although some factories derive their L-cysteine synthetically or from duck feathers, others get it from human hair. It's clean, though, thanks to the fact that the manufacturers who use human hair boil it in hydrochloric acid to extract the L-cysteine.

Yummy hair derivative aside, there's still the matter of that nasty strand stuck in the meatball. The FDA has set many standards for what it defines as "natural or unavoidable defects" in foods, but hair doesn't make the list. And if you think that's icky, there might be something even worse in your spaghetti; the FDA also okays up to two maggots per can of tomatoes.

WHAT'S THE DEAL WITH WISDOM TEETH ANYWAY?

▶ **THEY'RE A MOUTHFUL**

For human beings, the teeth known as the third molars—the last of a group of teeth that grinds food into easy-to-swallow chunks—tend to be overcrowded in adult mouths and thus require yanking. But every other toothed mammal has room for their "wisdom teeth," and so did Neanderthals and other early hominids, says evolutionary biologist Leslea Hlusko of the University of California, Berkeley. So why have those teeth become such a pain for us?

Genetics play a role in the shape and size of your jaw, but how it develops also depends on how much chewing stress you put on it during your childhood years, Hlusko says. Because we cook our food rather than tearing meat from the carcass, it's generally softer and easier to chew. As a result, over the generations our jaws have shrunk compared with those of our pre-agriculture ancestors, so we can't comfortably accommodate three sets of molars.

EAT YOUR VEGETABLES

There is some evidence that eating platefuls of raw carrots and other root vegetables as a kid might help your jaw grow large enough to hold wisdom teeth, but you're probably as well off crossing your fingers and hoping that you fall into the 15 percent of the population that never develops a set of third molars (and not one of the unlucky few who grow more than four). Of course, then you don't have an excuse to eat ice cream for a week.

▶ SURPLUS PARTS

Never. We're probably permanently stuck with our appendix, pinky toes, tailbone, and just about all of our other evolutionary holdovers. Wisdom teeth could eventually go, but significant changes like losing an appendage (teeth included) take millions and millions of years—who knows if humans will even be around that long. What's more, most of our seemingly useless vestiges are actually helpful.

The coccyx, or tailbone, "is an attachment point of a number of muscles at the pelvis. We need it for upright locomotion. It would be catastrophic if it went away," says Kenneth Saladin, an anatomist and physiologist at Georgia College and State University. The appendix, which helped our distant ancestors digest grass, has slowly evolved to take on a new purpose. Research led by William Parker and R. Randal Bollinger of Duke University has shown that the appendix now serves as a kind of "safe house" for the many microbes that aid in digestion. "Each of us has 900 to 1,600 species of bacteria in our gut to make sure we have a healthy immune system," says Stephen Stearns, an evolutionary-biology professor at Yale University. "If one takes over, or they all get flushed out by a disease, then the appendix functions like a holding tank for the good bacteria." Even the pinky toe helps keep our balance and diffuses impact throughout the foot when we run.

MAN-BOOBS

There are only a handful of truly useless parts of our body, but these are hanging on, too. As Saladin puts it, "Since vestiges like the muscles behind our ears have very little impact on reproductive success, there's no way to select against them." In other words, the ability to ear-wiggle doesn't interfere with having kids. Wisdom teeth were, like the appendix, good when we were eating lots of plants, but today only about 5 percent of us have jaws large enough for these extra molars. "Wisdom teeth are probably on their way out," Stearns says, "but it will take a long time."

The silliest of all vestiges is the male nipple. "Those don't have a function," Stearns says, "but they won't disappear, either." All embryos, male and female, begin developing according to the female body plan. Only around the sixth week of gestation do the genes on males' Y chromosomes kick in. "The developmental plan has the two nipples there, so you can't get rid of them genetically, because that would mess up the breasts of females." And nobody wants that.

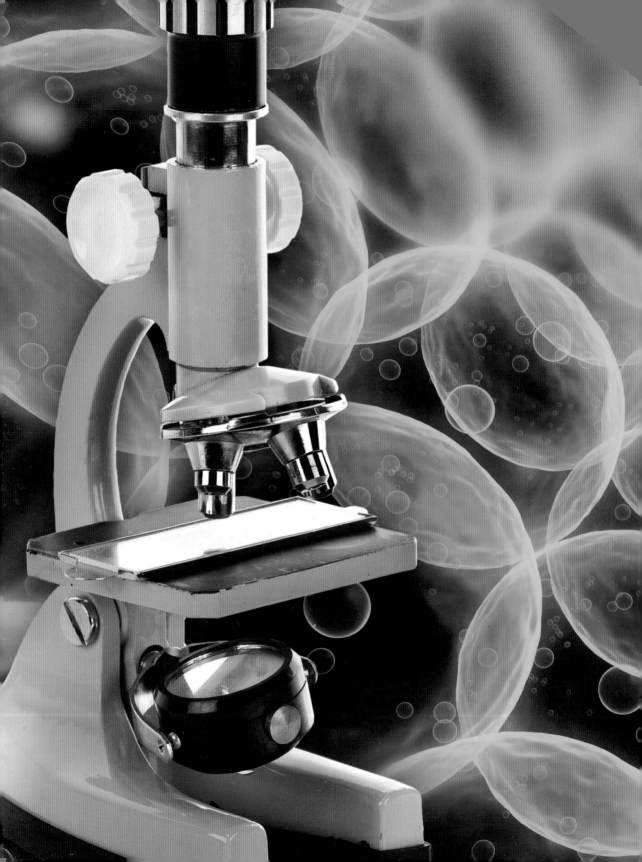

▶ **CELLULAR EVE**

Henrietta Lacks, a poor woman with a middle-school education who died of cancer in 1951, made one of the greatest medical contributions of all time. Her cells, taken from a cervical-cancer biopsy, became the first immortal human cell line—the cells reproduce infinitely in a lab. Although other immortal lines have since been established, Lacks's "HeLa" cells are the standard in labs around the world. Together they outweigh 100 Empire State Buildings and could circle the equator three times. *Popular Science* contributor Rebecca Skloot's book, *The Immortal Life of Henrietta Lacks*, tells the story behind the woman who revolutionized modern medicine. Here are five reasons we should all thank Henrietta Lacks.

1 Before HeLa cells, scientists spent more time trying to keep cells alive than performing actual research on the cells. An endless supply of HeLa cells freed up time for discovery.

2 In 1952, the worst year of the polio epidemic, HeLa cells were used to test the vaccine that protected millions from the disease.

3 Some cells in Lacks's tissue sample behaved differently from others. Scientists learned to isolate one specific cell, multiply it, and start a cell line. Isolating one cell and keeping it alive is the basic technique for cloning and in-vitro fertilization.

4 A scientist accidentally poured a chemical on a HeLa cell that spread out its tangled chromosomes. Later on, scientists used this technique to determine that humans have 46 chromosomes—23 pairs—not 48 as previously thought, which provided the basis for making several types of genetic diagnoses.

5 It was discovered that Lacks's cancerous cells used an enzyme called telomerase to repair their DNA, allowing them, and other types of cancer cells, to function when normal cells would have died. Anti-cancer drugs that work against this enzyme are currently in early clinical trials.

221

DOES A BLAND DIET REALLY HELP ALLEVIATE HEARTBURN?

▶ **RING OF FIRE**

Anecdotally, yes. Scientifically, no. Lauren Gerson, an associate professor of medicine at the Stanford University School of Medicine, says that none of the common dietary restrictions recommended to lessen heartburn—including limiting coffee, chocolate, caffeinated beverages, wine, and citrus fruits—have stood up to scientific scrutiny.

Heartburn occurs when the ring of muscle between the esophagus and the stomach that regulates food traffic fails to close properly. This can allow stomach contents, including stomach acid, to leak back into the esophagus, causing a burning sensation. Before the 1980s, when prescription "proton-pump inhibitors" such as Prilosec and, later, Nexium were introduced, patients had to rely on feeble antacids or even surgery. Diet restriction seemed to be the best noninvasive alternative to lessen the symptoms.

When Gerson's patients complained to her about their bland diets, she decided to go straight to the source—the original studies that implicated diet in the first place—to see if there was any support for the strict limitations imposed on heartburn sufferers. After slogging through more than 2,000 peer-reviewed journal articles, she concluded that there was not. Although some foods (like carbonated beverages) were shown to cause the muscle between the esophagus and stomach to relax—a potential cause of heartburn—Gerson says researchers never demonstrated whether consuming them would create the symptoms or if eliminating them from one's diet would lessen discomfort.

Blanket recommendations to cut out large classes of food may not be the answer to curing intractable heartburn, but old ideas about dietary limitations are not completely obsolete, says Kenneth R. DeVault, chair of the gastroenterology division at the Mayo Clinic in Jacksonville, Florida. "If something gives you symptoms," he says, "then you should probably avoid it." Chalk one up for anecdotal evidence.

RED HOT REQUIEM

There is no known case of a person dying from eating too many peppers, although several masochists have certainly tested the limits. The reigning king of jalapeño consumption is Chicagoan Patrick "Deep Dish" Bertoletti, who in 2010 set the International Federation of Competitive Eating record by downing 275 pickled jalapeños in a 10-minute time limit.

Looking to top Bertoletti and win a place in the Guinness Book of World Records is Anandita Dutta Tamuly, a woman from India who devoured 60 Bhut Jolokia peppers—one of the hottest peppers in the world—in just two minutes on national television. But she might not have anything on Mexico's Manuel Quiroz, who can squeeze habanero juice into his eyes without blinking.

But is there a deadly dose of spicy peppers? Researchers at Niigata University School of Medicine in Japan ran tests on mice to find out. After several hefty doses of pure capsaicin—the chemical that makes peppers hot—most of the mice died of lung failure. Don't worry, though—you'd have to eat hundreds of thousands of jalapeños to get the same dose, and the Bertolettis of the world notwithstanding, most people beg for mercy after a dozen.

223 CAN EATING HOT PEPPERS BE GOOD FOR YOU?

SPICE OF LIFE

Even if you don't have freakish tolerance levels, scientists have found that eating peppers can have medical benefits. A few years back, researchers at Cedars-Sinai Medical Center in Los Angeles found that capsaicin can kill human prostate-cancer cells grown in mice. The scientists estimated that the dosage was equal to a man weighing 200 pounds (91 kg) eating three to eight habanero peppers three times a week. So put some more five-alarm salsa on that burrito— it's not just good, it's (possibly) good for you.

224

HOW QUICKLY COULD A SINGLE **SUPERVIRUS** SPREAD TO **EVERY SINGLE** PERSON ON EARTH?

▶ EPIDEMIC PROPORTIONS

If it's a particularly contagious virus, it would spread across the planet in a year. "If it starts in New York, it's going to be in London certainly within a week," says Ira Longini, a biostatistician at the University of Washington and the Fred Hutchinson Cancer Research Center in Seattle who uses computer models to analyze how viruses globe-trot. "And from there, it will quickly travel to the rest of North America and Europe." For Longini's computer forecasts to become reality, though, certain conditions would need to be met.

First, it should be a strain of influenza. As anyone who has suffered through a bout of flu knows, it affects the respiratory tract, so sneezing and coughing make it easy to infect anyone within a radius of 3 feet (0.9 m). Second, the virus must originate in a major city with plenty of airport traffic, to ensure officials are off the trail of the real megabug, says Andrew Pekosz, a virologist and immunologist at Johns Hopkins University. The idea seems to freak him out. "With everybody expressing similar symptoms, we'd end up chasing, chasing, chasing, but always being a few steps behind, never really able to interrupt the spread."

225
ANTIBACTERIAL SOAP KILLS 99.9 PERCENT OF GERMS.
SHOULD I WORRY ABOUT THAT OTHER 0.1 PERCENT?

▶ **CLEAN FREAK**

Your dirty hands can harbor millions of germs, but simply washing your hands with regular soap—making sure you vigorously rub them together for 30 seconds—will slough enough microbes down the drain to cut that number to the tens of thousands. Assuming you don't then lick your hands, you're probably safe at this point, but there's still some risk. "Most pathogenic organisms cause disease when the numbers ingested are in the thousands to 10,000," says Dial Soap's manager of microbiology, George Fischler. Dial lab tests have shown that antibacterial soap, which most frequently uses the germ-killing agent triclosan, will, if used properly, reduce the number of germs on your hands to a few thousand. But Allison Aiello of the University of Michigan School of Public Health isn't convinced. Her lab, she says, has found no germ-killing benefit from triclosan over regular soap alone, even after three minutes of scrubbing.

If you're worried about runoff from antibacterial hand soap creating super-bacteria, your drainpipes are safe. Aiello and her colleagues have identified only a few bacterial strains resistant to triclosan or other household antibacterial products, and those were only in controlled laboratory settings primed for growing bacteria.

226
WHICH OF MY ORGANS COULD I SELL WITHOUT DYING?

DON'T TRY THIS AT HOME

First, a disclaimer: selling your organs is illegal in much of the world. It's also very dangerous. Handing off an organ is risky enough when done in a top hospital, even more so if you're doing it for cash in a back alley. No, really: don't do this. OK? OK.

There are many organs one can theoretically do without, or for which there's a backup. Most folks can spare a kidney, a portion of their liver, a lung, some intestines, and an eyeball, and still live a long life. That said, donating a lung, a piece of liver, or a section of intestines is a very complicated surgery, so it's not done frequently on the black market. And no one's going to make much cash on an eyeball. "In the U.S., there's a fairly steady supply of donated corneas from corpses," says Sean Fitzpatrick, director of public affairs at the New England Organ Bank. "There's pretty much no market demand for eyes." Giving up a kidney, though, is a relatively simple surgery that has netted desperate people a few bucks.

227
HOW MUCH CASH COULD I GET FOR MY ORGANS?

ORGAN AUCTION

Black-market organ dealers don't do a great job of filing taxes, but here are some prices based on rumored deals and reports from the World Health Organization. In India, a kidney fetches around $20,000. In China, buyers will pay $40,000 or more. A good, healthy kidney from Israel goes for $160,000.

Don't expect to pocket all that dough yourself, though. "The person giving up the organ only gets a fraction of the fee," says Sally Satel, a scholar at the American Enterprise Institute think tank who studies the prices paid by legal and illegal organ-donor operations. After the organ broker—the guy who sets up your kidney-for-cash transaction—takes his cut, he needs to pay for travel, the surgeon, medical supplies, and a few "look-the-other-way" payoffs. Most people get $1,000 to $10,000 for their kidney (probably much less than you were hoping for).

The best bet is to wait until compensation for organs is legalized in the U.S.—the Organ Trafficking Prohibition Act of 2009 would have allowed payment to donors, but it stalled in Congress—because there's certainly a market for kidneys. Recently, a man offering one of his for $100,000 (plus medical expenses) on Craigslist received several offers until the site removed his post. And you could probably hold out for even more. In 1999, before eBay delisted a kidney put up for auction, bidders drove the price up to $5.75 million.

▶ **SHARE AND SHARE ALIKE**

A few transplants out of the 28,000 or so performed every year in the United States involve the same organ spending time in more than two bodies. The most common scenario arises when a patient in the late stages of a disease receives a new liver or kidney as a last-ditch effort to keep him alive. If he dies shortly after, and the new organ wasn't the cause, retransplanting may be a slightly icky but viable option.

There are a few good reasons, however, why donated organs aren't often regifted. If the organ is coming from someone who was sick enough to need a new organ, it probably lived a rough second life. What's more, dying involves the entire body shutting down. "The trauma of dying can injure an organ," says Robert Montgomery, the director of the Johns Hopkins Comprehensive Transplant Center. "And then the second person dies, and the organ is taken out again. That's more injury." But the main problem with playing hot potato with an organ is the scar tissue that forms on it within weeks after the first surgery. Doctors must remove that tissue before a second transplant, and this can injure the organ too much to make it worth redonating.

But don't worry: Organs that are suitable for retransplantation rarely spend much time in the first recipient, which means less time for scar tissue to form. So if you're getting a third-hand kidney, chances are it's almost as good as almost new.

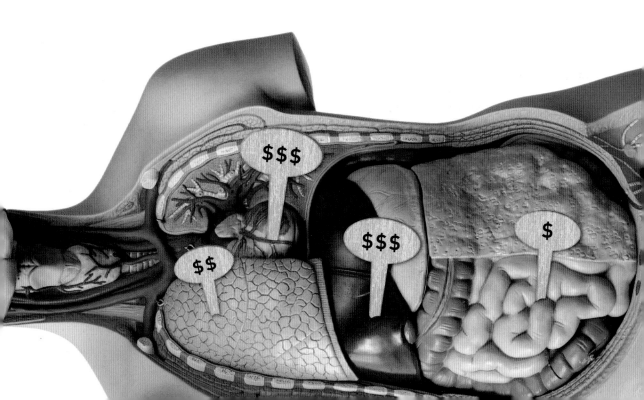

COULD SCIENTISTS REALLY CREATE A "ZOMBIE APOCALYPSE" VIRUS?

WHAT MAKES A ZOMBIE A ZOMBIE ANYWAY?

There's quite a variety of zombies out there. In West African and Haitian vodou, zombies are said to be human beings without a soul, their bodies nothing more than shells controlled by powerful sorcerers. In the 1968 film *Night of the Living Dead*, an army of shambling, slow-witted, cannibalistic corpses reanimated by radiation assault a group of rural Pennsylvanians. Let's say we're looking for something right in between Haiti and Hollywood: an infectious agent that will render its victims half-dead but still-living shells of their former selves. An effective agent would target, and shut down, specific parts of the brain, says Steven C. Schlozman, an assistant professor of psychiatry at Harvard University and author of *The Zombie Autopsies*, a series of fictional excerpts from the notebooks of "the last scientist sent to the United Nations Sanctuary for the study of ANSD," a zombie plague. Schlozman explains that although the walking dead would have some of their motor skills intact—walking, of course, but also the ripping and tearing necessary to devour human flesh—the frontal lobe, which is responsible for morality, planning, and inhibiting impulsive actions (like taking a bite out of someone), is nonexistent. The cerebellum, which controls coordination, is probably still there but not fully functional. This makes sense, since zombies in movies are usually easy to outrun or club with a baseball bat. The most likely culprit for this partially deteriorated brain situation, according to Schlozman, is as simple as a protein—specifically, a proteinaceous infectious particle, a prion.

WHAT'S A PRION?

Not quite a virus, and not even a living thing, prions are nearly impossible to destroy, and there's no known cure for the diseases they cause. The first famous prion epidemic was discovered in the early 1950s in Papua New Guinea, when some members of the Fore tribe were found to be suffering from a strange tremble. Occasionally a diseased Fore would burst into uncontrollable laughter. The tribe called the sickness "kuru," and by the early 1960s, doctors had traced its source back to the tribe's cannibalistic funeral practices, including brain-eating. Prions gained greater notoriety in the 1990s as the infectious agents that brought us bovine spongiform encephalopathy, also known as mad-cow disease. When a misshapen prion enters our system, as in mad cow, our mind develops holes like a sponge. Brain scans from those infected by prion-based diseases have been compared in appearance to a shotgun blast in the head.

EXCELLENT, SO LET'S DESTROY SOME BRAINS

Now, if we're thinking like evil geniuses set on global destruction, the trick is going to be attaching a prion to a virus. Prion diseases typically stay confined to a small population, so to make things truly apocalyptic, we need a virus that spreads quickly and will carry the prions to the frontal lobe and cerebellum. Targeting the infection to these areas is going to be difficult, but it's essential for creating the shuffling, empty-headed creature we expect. Jay Fishman, director of transplant infectious diseases at Massachusetts General Hospital in Boston, proposes using a virus that causes encephalitis, an inflammation of the brain's casing. Herpes would work and so would West Nile, but attaching a prion to a virus is, Fishman adds, "a fairly unlikely" scenario. And then, after infection, we need to stop the prion takeover so that our zombies don't go completely comatose, their minds rendered entirely useless. Schlozman suggests adding sodium bicarbonate to induce metabolic alkalosis, which raises the body's pH and makes it difficult for prions to proliferate. With alkalosis, he says, "you'd have seizures, twitching, and just look awful—like a zombie."

POPULAR SCIENCE
THE FUTURE NOW

ABOUT THE MAGAZINE

Founded in 1872, *Popular Science* is the world's largest science and technology magazine, with 6.7 million readers. Each month, *Popular Science* reports on the intersection of science and everyday life, delivering a look at the future now. It's the ultimate guide to what's new and what's next. *Popular Science* is one of the oldest continuously published magazines in the United States, and is published in five languages and nine countries. Its website, PopSci.com, has been readers' first stop for up-to-the-minute science news since the site first went live in 1999.

06323326300

▶ **ABOUT THE EDITOR**
Bjorn Carey edited *Popular Science*'s FYI column, where
he learned how to convince serious scientists to answer
silly questions. He is the managing editor of Life's Little
Mysteries, and an adjunct professor of journalism at New
York University. He has been a staff writer at LiveScience
and Space.com, and his work has been published in a
scientific journal. He studied biochemistry at Bowdoin
College, and earned a master's in science journalism from
NYU. He lives in New York City's West Village with his
wife, Corey, and son, Henry.

weldon**owen**

415 Jackson Street, Suite 200
San Francisco, CA 94111
Telephone: 415 291 0100
Fax: 415 291 8841

www.wopublishing.com

A division of

BONNIER

Library of Congress Control Number: 2011925881

ISBN: 978-1-61628-120-5

Printed in China by Toppan Leefung Printing Limited

10 9 8 7 6 5 4 3 2

WELDON OWEN INC.

CEO, President Terry Newell

**VP, Sales and New Business
Development** Amy Kaneko

VP, Publisher Roger Shaw

Creative Director Kelly Booth

Executive Editor Mariah Bear

Editor Lucie Parker

Project Editor Katharine Moore

Senior Designer Stephanie Tang

Image Coordinator Conor Buckley

Production Director Chris Hemesath

Production Manager Michelle Duggan

Color Manager Teri Bell

PHOTO CREDITS

All photographs courtesy of Shutterstock except:
CERN PhotoLab: Macimilien Brice 16; NASA 45, 50; Corbis: 148;
WikiCommons 131, 153; Alexander Ivanov 157; iStockphoto 198, 202

Image treatment and photo collaging: Conor Buckley,
Scott Erwert, Stephanie Tang

Illustrations: Conor Buckley 3, 42, 138, 156

ARTICLE CONTRIBUTORS

All articles written by Bjorn Carey except:
Lauren Aaronson 38; Natalie Avon 62; Julie Beck 64, 65; Corey Binns
61, 72, 73, 108, 109, 116, 117, 118, 165, 172, 173, 185, 186, 211; Lana
Birbrair 42, 169, 170, 228; Ryan Bradley 229; Michelle Bryner 204;
Darrin Burgess 76, 151; Alessandra Calderin 15, 45, 137; Melissa A.
Calderone 21, 22, 23, 60, 181, 182; Doug Cantor 162, 163; Jessica
Cheng 43, 91, 148; Matthew Cokeley 152, 153, 199; Jonathan
Coulton 57, 58; Stuart Fox 4, 8, 18, 189, 190; Danny Freedman 16, 17,
19, 20, 39, 114, 115, 135, 191; Amy Geppert 46, 149, 166, 225; Day
Greenberg 133; Rachel Horn 71, 131, 207; Cornelius Howland 160;
Jeremy Hsu 183, 184; Fred Koschmann 78, 79; Abigail W. Leonard
90, 175; Susannah F. Locke 44, 51, 155; Amanda Macmillan 99, 210;
Eric Mika 41, 102, 203; Brandon Miller 56, 200, 222, 223; Christopher
Mims 5, 89, 95, 139, 154; Gregory Mone 150; Todd Neale 28, 47,
48, 59; Craig Nelson 50; James Norton 27; Holly Otterbein 132, 140,
141, 144; Rosa Pastore 224; Nicole Price-Fasig 25, 26, 63, 136, 178,
179, 221; Sandeep Ravindran 11, 12, 159; James Riordon 86, 87, 88;
Michael Rosenwald 67; David Rothenberg 125; Katherine Ryder 13;
Amber Sasse 174, 218; Lizzie Schiffman 37; Amanda Schupak 214,
215; Catherine Schwanke 92, 113, 124, 161; Abby Seiff 52, 53, 68, 69,
96, 97, 110, 142, 164, 171; Graeme Stemp-Morlock 176, 177; Carina
Storrs 35, 85, 196; Dawn Stover 121; Elizabeth Svoboda 49, 77, 94,
145, 146; Ker Than 7, Carla Thomas 143; Kalee Thompson 187, 188;
Jen Trolio 70; Melinda Wenner Moyer 84, 93, 197; Natalie Wolchover
74, 157, 219; Anne Wootton 33, 34, 75, 167; Sally Younger 6, 36;
Victor Zapana 216, 217

SPECIAL THANKS

Editorial and research support: Kathryn Ferguson, Emelie Griffin,
Jan Hughes, Robert F. James, Marianna Monaco, Michael Shannon,
Valerie Witte, Charlie Wormhoudt, Mary Zhang.

Original design concepts: Scott Erwert